Demand-Response in Smart Buildings

Demand-Response in Smart Buildings

Special Issue Editors

Denia Kolokotsa
Gloria Pignatta
Kostas Gobakis

MDPI • Basel • Beijing • Wuhan • Barcelona • Belgrade • Manchester • Tokyo • Cluj • Tianjin

Special Issue Editors

Denia Kolokotsa
Technical University of Crete
Greece

Gloria Pignatta
UNSW Sydney
Australia

Kostas Gobakis
Technical University of Crete
Greece

Editorial Office
MDPI
St. Alban-Anlage 66
4052 Basel, Switzerland

This is a reprint of articles from the Special Issue published online in the open access journal *Energies* (ISSN 1996-1073) (available at: https://www.mdpi.com/journal/energies/special_issues/smart_buildings).

For citation purposes, cite each article independently as indicated on the article page online and as indicated below:

LastName, A.A.; LastName, B.B.; LastName, C.C. Article Title. *Journal Name* **Year**, *Article Number*, Page Range.

ISBN 978-3-03928-266-1 (Pbk)
ISBN 978-3-03928-267-8 (PDF)

Contents

About the Special Issue Editors

Denia Kolokotsa is Associate Professor at the School of Environmental Engineering of the Technical University of Crete, Greece. Her research interests include energy management for the built environment, energy efficiency and renewables, integration of advanced energy technologies, indoor environmental quality, energy efficiency in buildings, cool materials, distributed energy management systems, artificial intelligence, building automation as well as optimisation of microgrids and smart grids. She is Editor of the *Energy and Buildings* journal of Elsevier and has served as Editor in Chief of *Advances in Building Energy Research*, Taylor and Francis. She is a member of the editorial board of 3 journals and Scientific Coordinator of 8 international research projects. She has participated in more than 30 international research projects as team leader or researcher. She is the author of 95 articles published in peer-reviewed journals. She is also President and Founding Member of the European Cool Roofs Council and member of numerous scientific committees and international organisations.

Gloria Pignatta is an academic of the Faculty of Built Environment at the University of New South Wales (UNSW Sydney), Australia. Her research touches on the global problem of climate adaptation and mitigation by investigating the relationship between extreme weather events and the built environment. She is a Civil Engineer with a Ph.D. in Energy Engineering from the University of Perugia, Italy. In 2016, she was Postdoctoral Fellow at the Inter-University Research Center on Pollution and Environment "Mauro Felli" (CIRIAF), University of Perugia, Italy. In 2017, she was Postdoctoral Associate at Singapore-MIT Alliance for Research and Technology (SMART), in Singapore. She is the author of scientific papers in the field of building energy efficiency and sustainability, and has participated in 2 European and several international research projects as team leader or researcher.

Kostas Gobakis is Research Associate at the School of Environmental Engineering of the Technical University of Crete, Greece. His research interests include prediction using artificial neural networks, energy simulation energy, management for the built environment, energy efficiency and renewables, energy efficiency in buildings, and cool materials. He is a physicist with a Ph.D. in Environmental Engineering from Technical University of Crete. He has served as the main technical coordinator of more than 5 European funded projects since 2012.

Preface to "Demand-Response in Smart Buildings"

Smart buildings and communities are currently in the spotlight of the scientific community due mainly to their potential for creating a more sustainable and livable future through the application of smart energy management. Demand response (DR) offers the capability of applying changes in the energy usage of consumers—from their normal consumption patterns—in response to changes in energy pricing over time. This leads to lower energy demand during peak hours or during periods when an electricity grid's reliability is put at risk. Therefore, demand response is a reduction in demand designed to reduce peak load or avoid system emergencies. Hence, demand response can be more cost-effective than adding generation capabilities to meet the peak and/or occasional demand spikes. The underlying objective of Demand response is to actively engage customers in modifying their consumption in response to pricing signals. Demand response is expected to increase energy market efficiency and the security of supply, which will ultimately benefit customers by way of options for managing their electricity costs, leading to reduced environmental impact.

This Special Issue entitled "Demand-Response in Smart Buildings" was published in *Energies*, and includes 5 international scientific contributions comprising 3 research papers and 2 review papers. Studies on energy retrofitting, consumer to fog to cloud framework, genetic algorithm optimisation models, energy policy, and energy legislations are collected and presented in this book. We would like to express our gratitude to all the authors and reviewers who have significantly contributed to this Special Issue. A special thanks also goes to the editorial team of *Energies* (MDPI) for offering us the opportunity to publish this book, especially Mr. Mark Guo, Senior Assistant Editor of *Energies*.

This book represents the Special Issue of *Energies*, entitled "Demand-Response in Smart Buildings", that was published in the section "Energy and Buildings". This Special Issue is a collection of original scientific contributions and review papers that deal with smart buildings and communities.

Demand response (DR) offers the capability to apply changes in the energy usage of consumers—from their normal consumption patterns—in response to changes in energy pricing over time. This leads to a lower energy demand during peak hours or during periods when an electricity grid's reliability is put at risk. Therefore, demand response is a reduction in demand designed to reduce peak load or avoid system emergencies. Hence, demand response can be more cost-effective than adding generation capabilities to meet the peak and/or occasional demand spikes. The underlying objective of DR is to actively engage customers in modifying their consumption in response to pricing signals. Demand response is expected to increase energy market efficiency and the security of supply, which will ultimately benefit customers by way of options for managing their electricity costs leading to reduced environmental impact.

Denia Kolokotsa, Gloria Pignatta, Kostas Gobakis
Special Issue Editors

Article

Intelligent Resource Allocation in Residential Buildings Using Consumer to Fog to Cloud Based Framework[†]

Sakeena Javaid [1], Nadeem Javaid [1,*], Tanzila Saba [2], Zahid Wadud [3], Amjad Rehman [4] and Abdul Haseeb [5]

[1] Department of Computer Science, COMSATS University Islamabad, Islamabad 44000, Pakistan; sakeenajavaid@gmail.com
[2] College of Computer and Information Sciences, Prince Sultan University, Riyadh 11586, Saudi Arabia; tsaba@psu.edu.sa
[3] Department of Computer System Engineering, University of Engineering and Technology, Peshawar 25000, Pakistan; zahidmufti@nwfpuet.edu.pk
[4] College of Computer and Information Systems, Al Yamamah University, Riyadh 11512, Saudi Arabia; drrehman70@gmail.com
[5] Department of Electrical Engineering, Institute of Space Technology (IST), Islamabad 44000, Pakistan; haseeb_karak@yahoo.com
* Correspondence: nadeemjavaidqau@gmail.com; Tel.: +92-300-579-2728
† This paper is an extended version of paper accepted in 14th IEEE International Wireless Communications and Mobile Computing Conference (IWCMC-2018), 2018 in Cyprus.

Received: 29 January 2019; Accepted: 21 February 2019; Published: 1 March 2019

Abstract: In this work, a new orchestration of Consumer to Fog to Cloud (C2F2C) based framework is proposed for efficiently managing the resources in residential buildings. C2F2C is a three layered framework consisting of cloud layer, fog layer and consumer layer. Cloud layer deals with on-demand delivery of the consumer's demands. Resource management is intelligently done through the fog layer because it reduces the latency and enhances the reliability of cloud. Consumer layer is based on the residential users and their electricity demands from the six regions of the world. These regions are categorized on the bases of the continents. Two control parameters are considered: clusters of buildings and load requests, whereas four performance parameters are considered: Request Per Hour (RPH), Response Time (RT), Processing Time (PT) and cost in terms of Virtual Machines (VMs), Microgrids (MGs) and data transfer. These parameters are analysed by the round robin algorithm, equally spread current execution algorithm and our proposed algorithm shortest job first. Two scenarios are used in the simulations: resource allocation using MGs and resource allocation using MGs and power storage devices for checking the effectiveness of the proposed work. The simulation results of the proposed technique show that it has outperformed the previous techniques in terms of the above-mentioned parameters. There exists a tradeoff in the PT and RT as compared to cost of VM, MG and data transfer.

Keywords: smart grid; requests time; cloud computing; energy management; response time; processing time; resource allocation; microgrid; fog computing

---

1. Introduction

In Smart Grig (SG), Demand Side Management (DSM) with the integration of Information and Communication Technologies (ICTs) is considered as its paramount function. Different electric appliances and control services are scheduled and integrated with DSM. These appliances and services are: shiftable loads, charging and discharging of the electric vehicles, smart devices (i.e., smart meters,

Distributed Generators (DGs)), etc. With the development of the massive electricity market, multiple entities are involved on the DSM side: NASDAQ trading organizations of OPower, C_3 Energy, etc. [1–3]. In order to optimize the energy management on demand side, these organizations apply existing techniques for the bidirectional interactions and online processing facilities. Multiple small businesses and standalone buildings are also contibuting in the electricity market on DSM in order to engage themselves in the development of the SG applications [4].

There are two significant aspects which are required to be considered in future DSM based on the above-mentioned scenarios. These aspects are the technical aspect and the economical aspect. The technical aspect considers the huge size data of the appliances which means the number of appliances used in the smart buildings, their power ratings, their On and Off status, and scheduling horizons. It needs to be processed by considering the specific time constraints for maintaining its computational complexity. Computational complexity consists of the resources which are required for executing the consumers' requests. The time required to respond to the requests is considered as Processing Time (PT) in our scenario and it is maintained by fulfilling the consumers' comfort preferences. Secondly, the economical aspect focuses on most of the newly developed buildings and businesses which are not participating in the ICT framework in new stages. It becomes complex to maintain its reliability without their participation in this case. Hence, allocating the ICT facilities: processing power, storage capacity and availability of the resources are the critical problems [5,6]. A flow diagram of cloud computing to its entities is shown in Figure 1.

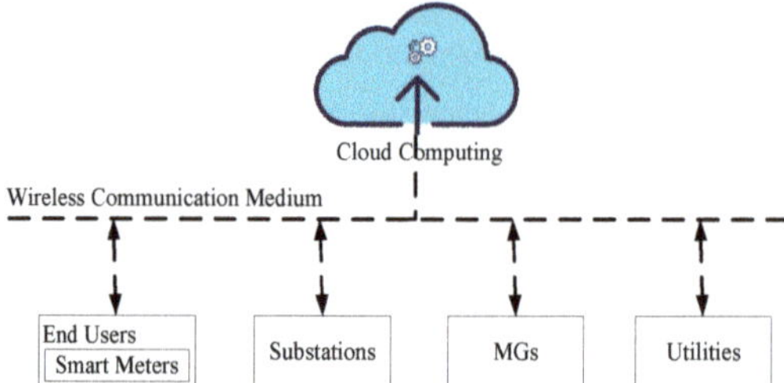

Figure 1. Flow Diagram of Cloud, Utilities, Consumers, Substations and Microgrids (MGs).

The management of computation, storage and on-demand resource availability from grid and electricity scheduling for the consumers is handled by cloud computing. Cloud computing resolves both technical and economical aspects of these (computation, storage and on-demand resource availability) problems. It is also the enhanced version of the parallel and grid computing. Fog computing is the specialized model of the cloud computing which is responsible for efficient management of the consumers' resources on the edge of the network. It upgrades the locality, reliability, security and latency of the consumers' demands [7–9].

1.1. Motivation

Previous studies [10–12] have incorporated the cloud services in SG environment. Cloud computing based infrastructure has been presented for the future generation power grid by Luo et al. in [10]. Authors have considered the charging and discharging schedules of the electric vehicles in a decentralized manner to procure the load shuffling facility [11]. The scheduling problem is formulated with the help of mixed discrete programming technique. However, they have not incorporated the cloud platforms by organizing the demands of electric vehicles' customers. Another decentralized algorithm for optimally scheduling the electric vehicles charging has been proposed by Gan et al. [12] and the proposed algorithm has led to the exploitations in terms of electric vehicles loads for filling the valleys in their load profiles. Fog computing [13] is integrated as the middle layer between the consumers and cloud environment. It minimizes the latency and improves reliability of the cloud services. All of the aforementioned studies are either based on the cloud or fog based environment. None of the earlier techniques are based on the fog and cloud environment together for the optimal resource allocation. Although, cloud provides on-demand availability of the resources; however, it degrades the latency and effects Response Time (RT) for the consumers which creates frustrations. Fog computing resolves these issues very efficiently. We have proposed the concept of C2F2C based environment in order to minimize the latency and efficient scheduling of the resources in the residential buildings. This study also considers the optimal RT, cost of data transfer, MG and Virtual Machines (VMs), consumers' requests time and PT, which is provided by fog computing.

1.2. Contributions

In this work (an extension of [14]), we have proposed and implemented the Consumer to Fog to Cloud (C2F2C) based framework for efficient resource management in residential buildings. Our main contributions are described as below:

- The C2F2C based framework is presented for efficient management of the consumers' demands in the residential buildings.
- Six regions of the world are considered for optimal resource allocation of residential users using the fog layer as the middle layer. Fog layer stores the consumers' SM information on each local region's fog for the consumers.
- MGs are integrated in order to fulfill the consumers' energy requirements while minimizing the total cost for the end users.
- RT, Requests Per Hour (RPH), PT and cost are optimized using the proposed: Shortest Job First (SJF) algorithm. This is also compared with the previous two algorithms: Round Robin (RR) and Equally Spread Current Execution (ESCE). From simulation results, it has been observed that SJF outperformed the RR and ESCE in terms of the aforementioned parameters.
- Simulations are conducted for two scenarios to validate the effectiveness of the proposed framework: (1) resource allocation using MGs and (2) resource allocation using MGs, electric vehicles and power storage devices.

The remainder of the manuscript is summarized as follows. In Section 2, literature review is discussed along with the existing limitations. Section 3 describes the proposed system model and simulation results are elaborated in Section 4. Finally, the conclusion and future challenges are described in Section 5.

2. Literature Review

In order to subsist the intensive demand of the electricity, different methodologies have been studied from existing work. These methodologies are categorized based on their proposed architectures: (1) methodologies for cloud based architecture and (2) methodologies for fog based architecture. So, the techniques for energy management in SG using cloud computing have been discussed first then the techniques for energy management in SG using fog computing have been discussed.

2.1. Methodologies Regarding Cloud Based Architecture

A novel architecture for electric vehicles' charging and discharging is presented using public supply stations [15]. For electric vehicles' charging and discharging, two priority assignment algorithms: (1) random priority attribution and (b) calender priority attribution are developed in the cloud environment. These algorithms are used for monitoring the waiting time of electric vehicles in order to maintain grid stability in peak hours by setting the demand supply graph as a constraint. Smart phone has been considered as a component of the cyberphysical system for dynamic voltage scaling by minimizing the frequency of smart phone which leads to energy minimization [16]. Authors also develop an energy aware dynamic task scheduling algorithm for minimizing the aggregated energy requirements of the running applications by considering two constraints: (1) time and (2) probability. This work has been restricted to the energy minimization in the smart phone system. It also lacks its applicability in SG energy management domain.

In [17], one new communication model is proposed which is composed from two models: the cloud based Demand Response (DR) model and distributed DR model for optimizing the communication delay. This model suffers from huge cost for the peak demand scenarios. Further, a cost computation model is also proposed for demand side consumers by the optimal use of the cloud resources. Moreover, this model provides an opportunity for the consumers regarding optimized resource availability which helps in cost minimization. They have designed the modified priority list algorithm which is used for optimal distribution of the cloud resources (instances): on-demand instances and reserved instances. It minimizes the total operational cost of the system.

The idea of nanogrids is presented in sustainable buildings for multi-tenant cloud environment in [18]. A game theory technique using the collaboration for power management in SG cyberphysical system is proposed in [19], where payoff function is formulated by investigating each player's information (i.e., transmission and service delay information) through conditional entropy. Moreover, the dynamic workflow management is designed for executing the tasks via virtual cloud platform. Various VMs are utilized in a coalition for running the jobs in an intelligent way for this scenario. In [20], the energy hubs are manoeuvred for storage of the consumers' data in the cloud environment for efficient DSM. Stochastic dynamic programming is used to manage the load in real time environment for reducing the expenses by incorporating the consumers' participation. This system is particularly designed for cloud environment and our system incorporates the fog in order to minimize the latency, PT and cost.

Some reviews about cloud computing technologies are presented in [13] along with their challenges. Some limitations of the existing power systems are also described in this work by the cloud computing technologies, for instance; the infrastructure of the power system is weak and there can be blackouts at any time due to uncertain catastrophes. For maintaining the resiliency, cloud computing facilitates the consumers by providing computing, storage and security features.

2.2. Methodologies Regarding Fog Based Architecture

In [21], authors propose a cloud based DSM system using MGs which manages energy for the consumers from multiple regions in order to minimize the utility and consumers' cost. This technique also reduces the time and efforts by incorporating the modularity feature for developing smart cities. Bi-level optimization algorithm using linear cost function is also developed and applied in this scheme. Authors in [22] propose energy management scheme for electricity and natural gas network using integrated DSM. This scheme differentiates the electricity consumption techniques because they consider the consumption regarding each user by considering multiple users' interactions. Nash Equilibrium (NE) is used in this technique for measuring the interactions of the players. NE is applied for electricity cost and peak load reduction.

Authors present a new SG architecture for electric vehicles' scheduling in buildings (integrated with MGs) which enable a central SG controller for cloud Data Center (DC) environment [23]. This system helps the SG users to optimize the load requests in an efficient fashion. It enhances the stability of grid during the high demand intervals. Cloud computing helps the consumers by providing: computing, storage and networking capacities. In addition, there are also some limitations of cloud, i.e., latency, security and downtime problems. In order to ameliorate the latency, reliability, and resiliency of the services, one new strategy is presented in [24], in which fog platform is utilized by considering the energy management as a service. This strategy also discusses two energy management prototypes: home energy management prototype and MG energy management prototype for electricity bill and delay minimization. Security for the customers' private data is enhanced through encryption, which improves the flexibility and reliability by augmenting the fog computing principles [25,26]. It is also responsible for providing the locality of consumers' appliances.

Afterwards, a new approach regarding electricity cost reduction problem presents the internet payload requests and SG dynamic electricity pricing mechanism [27]. This approach also discusses the predictive cost control for the smart charging on both sides: (1) battery energy for the servers and (2) electricity from the power grid. In order to mitigate the overall cost of the system, batteries are charged during the low price rate hours and are discharged during the high price rate hours. In [28], authors use big data analytics for balancing the energy of the residential, industrial and commercial buildings. Multiple big data anaytics and decision making functions are used for efficiently managing the energy in these buildings. This work also considers the comfort preferences of the consumers using the control configuration and planning processes. This technique is designed for the buildings of smart cities.

The aforementioned schemes are described for the load scheduling of appliances in the buildings. However, integrated cloud and fog based scheme is not described for optimized resource allocation in any of the regions of the world. The proposed work describes the energy management in the residential buildings using C2F2C framework.

3. Proposed System

The C2F2C framework is proposed for intelligently managing the resources regarding the electricity demands of the residential users. In the subsequent subsections, assumptions of the system model, problem formulation and proposed C2F2C framework are described.

3.1. Assumptions

In the proposed work, the following six assumptions have been taken into account.

- The world is categorized into six regions on the basis of six continents.
- In every region, there is a cluster of buildings considered for the optimal resource allocation.
- Integrated cloud and fog environment has been used for optimizing the consumers' demands.
- Residential users are facilitated with the required services (electric power supply) despite the limitations of the cloud. Fogs and MGs are used for responding to the consumers' requests (i.e., by omitting the latency issues with the integration of fog layer).
- Every region has one fog server for efficient resource management.
- For high computation, storage and resource management, VMs are assigned with the DC resources.

3.2. Problem Formulation

We have considered the following components for resource allocation in the residential buildings all over the world which are categorized as: set of regions, buildings, VMs, MGs and DCs. Let VM indicate the set of VMs and each of them processes the tasks T, so the sets of the VM and tasks T are mathematically represented by Equations (1) and (2):

$$VM = \{vm_1, vm_2, ..., vm_n\},\tag{1}$$

$$T = \{t_1, t_2, ..., t_n\},\tag{2}$$

here, the set of the VMs $(vm_1, vm_2, ..., vm_n)$ and tasks $(t_1, t_2, ..., t_n)$ are between 1 to n. In this environment, all VMs are working in parallel with each other and have the same capacity. The tasks are executed in a non-preemptive fashion on each VM. Through hypervisor, every VM is monitored and it is assigned to certain tasks, if they are required to be processed. The total tasks assigned to each VM are dependent on the number of consumers' requests and their PT, RPH and RT. Each of the aforementioned parameters are formulated using the linear programming [29]. Equation (5) is computed for the PT using the Equations (3) and (4). Equation (3) represents the tasks and VMs assignment states.

$$Tasg_{kl} = \begin{cases} a; & for\ assigning\ the\ tasks, \\ b; & for\ assigning\ the\ VMs. \end{cases} \tag{3}$$

where, "a" indicates the set of the VMs and "b" denotes the set of tasks. $Tasg_{kl}$ is used to indicate the assignment of VMs and tasks to each VM. The RPH is evaluated by Equation (4).

$$RPH_{kl} = \begin{cases} 1; & if\ task\ is\ assigned, \\ 0; & if\ task\ is\ not\ assigned. \end{cases} \tag{4}$$

RPH_{kl} indicates the number of VMs assigned to the tasks requested by each consumer in the C2F2C system. After finalizing the RPH and the tasks assignment to each VM in the system, its total PT (PT_{tot}) can be computed by Equation (5);

$$PT_{tot} = \Sigma_{k=1}^{n}\Sigma_{l=1}^{m}(PT_{kl} * RPH_{kl}). \tag{5}$$

PT_{kl} represents the PT of the VMs assigned to each task and RPH_{kl} indicates the RPH of every VM dedicated to the assigned task. The RT is computed with the help of Equation (6) as given below:

$$RT_k = \frac{\Sigma_{k \in VMs}(CT_k)}{makesspan * Number\ of\ the\ VMs}. \tag{6}$$

Here, RT_k is used for denoting the total RT of the VMs in the proposed system and CT_k is the completion time of the tasks. "Makesspan" is a term which describes the aggregated running time of the tasks. The cost of the system is computed via aggregating the VMs cost, MGs cost and data transfer cost. VM cost is computed on the basis of the Multiple Instructions Per Second (MIPS) and the request size, whereas the data transfer cost is computed by considering the total data and the assigned bandwidth to that data. In addition, MG cost is computed based on the number of the installed MGs. So, the total cost of the system is computed by Equation (7):

$$Cost_{tot} = Cost_{Datatrans} + Cost_{VM} + Cost_{MG}. \tag{7}$$

Here, the variables: $Cost_{tot}$, $Cost_{Datatrans}$, $Cost_{VM}$ and $Cost_{MG}$ used for cost in Equation (7) represent the total cost, data transfer cost, VM cost and MG cost, respectively.

3.3. Proposed C2F2C Framework

In this paper, a C2F2C framework is proposed for resource allocation in residential areas as shown in Figure 2. The proposed framework is based on the following resources: electricity and its requested entertaining facilities like MG, cloud and fog for smart buildings. Cloud and fog also have some resources such as computation, storage and networking facilities (between the end devices to the service providers). These resources are responsible for managing the demands of consumers in six regions of the world. The whole framework is comprised of a cloud, six fogs within the regions, set of buildings, set of homes, and their load requests. These six regions are considered for the residential energy consumption and management. Each region has one fog for fulfilling its demands. When clusters of the buildings send load requests to the fog server for completing the load demands, this framework uses MGs on first priority to fulfill the load demands of the consumers, otherwise, it communicates to the cloud for exchanging services with the utilities. There are two MGs which are deployed in each region as shown in Figure 2. MGs are based on distributed generation, loads, storage devices, etc. and integrated with the residential buildings. In each region, MGs are installed by the users and managed by the fogs. The proposed framework acts as the automation of the energy management services and requests through fog environment.

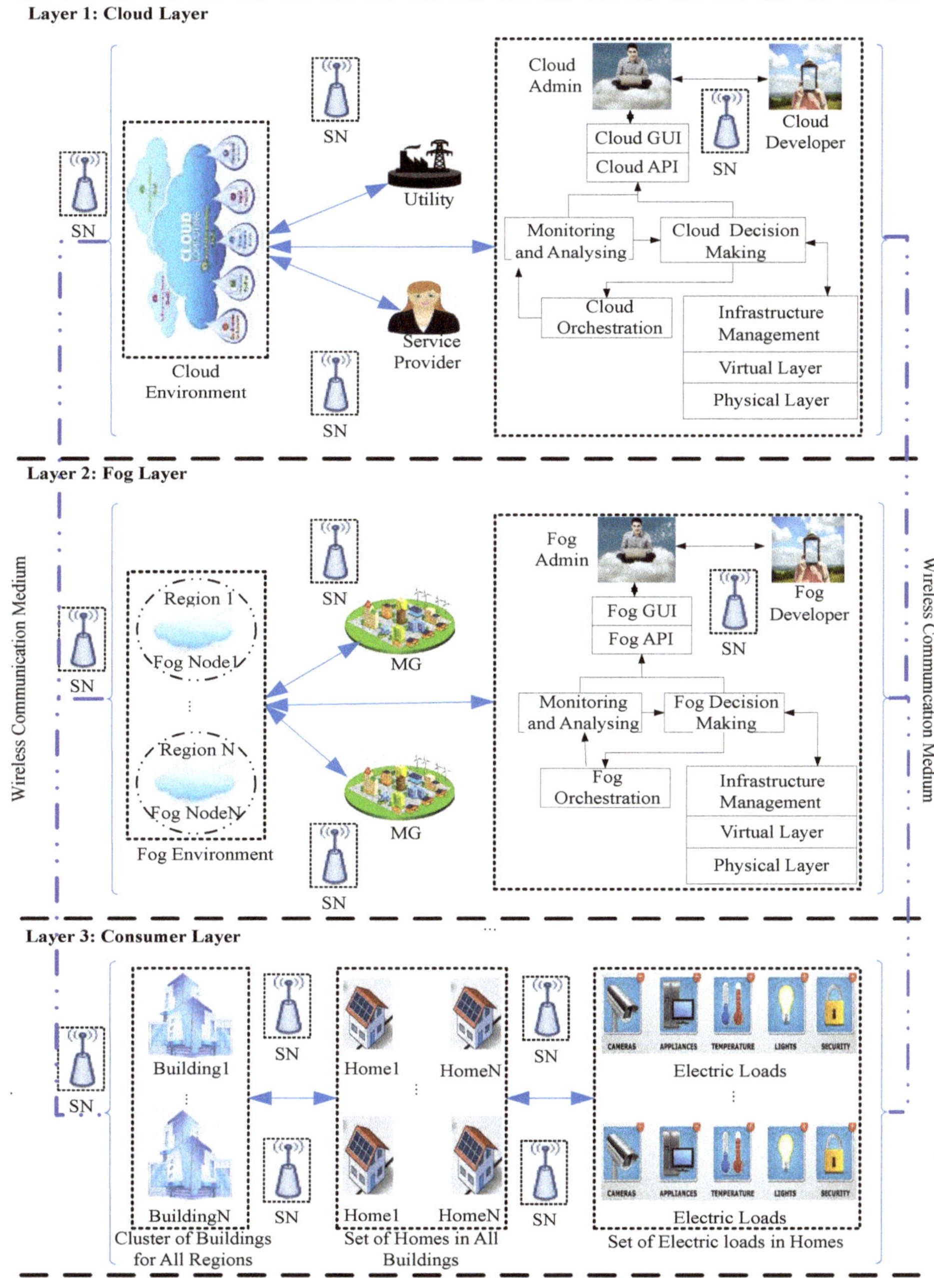

Figure 2. C2F2C based Framework, Components and Communication Process.

The proposed framework is comprised of three layers: consumers' layer, fog layer and the cloud layer. Cloud and fog layers are used for orchestrating and controlling the cloud and fog resources in consideration of the consumers' requests. Consumers utilize cloud and fog facilities for fulfilling their daily load requirements. Both cloud and fog layers show the cloud and fog admin and developers for maintaining their internal resources as shown in Figure 2. They maintain their services effectively according to the consumers' demands. Data is transferred to the cloud when fog's resources are completely utilized because the fog has limited resources. Fog only communicates to the cloud when all of its resources are utilized and there is an excess demand from the consumers of any region. When requests are sent to the cloud, it does not resolve any of the users' cost and latency issues; however, for resolving the limited resources unavailability, cloud is utilized. Storage resources are not considered in this work. The third layer depicts the set of buildings, number of homes and their load demands in each region.

The communication medium is used as the wi-fi for maintaining the communication among all C2C2F layers and their respective components. When any home in the buildings sends requests, it is first received at the fog server, then the availability of resource is checked. After verifying the availability of the requested resource, it is processed through VMs and is also conveyed to the consumers. Hypervisor is used to monitor the task execution states of VMs at the cloud end as shown in Figure 3. Each fog is responsible for its own region's services. The region may be any continent which lies in the communication range of that fog. The following kinds of the communications are used in this scenario: fog to cloud, consumer to fog, appliance to appliance and vise versa. Sensors Nodes (SNs) are used in each layer for sensing and sharing the information among components.

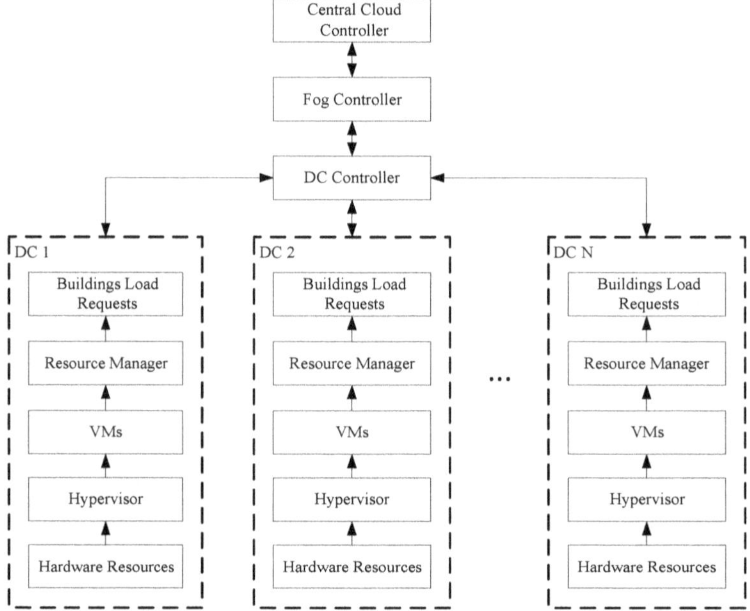

Figure 3. Resource Management in Buildings, Cloud and Fog's DC.

There are three types of consumers who are considered in this framework: traditional consumers, consumers with the Home Energy Management System (HEMS) and the consumers who are integrated with the local generation and HEMS. Traditional consumers do not have HEMS or any local generation whereas the consumers that have HEMS and the on-demand cloud services are referred to as smart consumers. In addition, the consumers having both HEMS and the local generation are considered as prosumers and the smart users in the buildings. Traditional consumers do not participate in the energy management programs.

Figure 3 shows the resource management in the cloud and fog DCs where VMs are used as the resource manager and hypervisor is allocated to monitor the VMs according to their existing status. DCs are considered variable in this work. The required hardware resources for each DC are based on fog specification which are elaborated in Table 1. Two scenarios are used for the resource optimization using this framework. Each scenario is further comprised of two scenarios. One scenario considers the 25 VMs for resource allocation with single DC whereas the second scenario considers the 50 VMs with two DCs for the optimization of the resources in the residential buildings. The extended scenarios are elaborated in the subsequent subsections.

Table 1. Input Parameters for both scenarios.

Parameters	Values
Number of VMs	25, 50
Executable Task size	250
Users Grouping Factors	1000
Number of DCs	1, 2
DC Processor Speed	100 MIPS
VM Image Size	10,000 GB
DC Available BW per Machine	10,000
DC Storage per Machine	100,000 Mb
DC Memory per Machine	2048 Mb
DC Number of Processors per Machine	4
DC Architecture	X86
VM Memory	1024 Mb
VM Bandwidth	1000 Mb
DC VMM	Xen
DC VM Policy	Time Shared
Requests Grouping Factors	100

3.4. Twenty Five VMs with Single DC

In this scenario, 25 VMs are considered with the single DC in order to fulfill the load demands of the consumers from every region of the world. Load demands are assigned to the VMs which are responsible for the management of these demands and services. Each VM is monitored by the hypervisor (as discussed earlier) which decides that either it has finished its task or it is in idle state. Every region's load demands are evaluated using this scenario for validating the proposed framework. Cloud and fog communicate with each other in order to manage the resources whereas fog and consumers interact for assigning the facilities to the requests of consumers. Consumers send requests to each fog in their region from the buildings. Residential buildings in each cluster vary between 80 and 150. The input parameters and their values for the scenario 1 and 2 are given in Table 1 [30].

3.5. Fifty VMs with Two DCs

Scenario two is also running with the same simulation setup as in scenario 1. The set of the DCs and VMs are varied for the maximum resource utilization. This scenario integrates the 50 VMs with two DCs. MGs are considered to be the same as the above scenario and communication mechanism of the entities is also similar to scenario 1: consumer to fog, fog to cloud, appliance to appliance and vise versa. In addition, scenario 2 also comprises on the power storage devices and on electric vehicles as described in detail in the simulation section.

Three algorithms have been used for resource allocation in the C2F2C environment: ESCE, RR and SJF algorithms. These algorithms are applied for assignments of the VMs, MGs and other resources in the DCs for the consumers in any region as described in the Algorithms 1–3. When requests are received from the residential buildings, ESCE algorithm first checks the resources and assigns them to the available VMs. RR receives all the requests in an organized (cyclic) fashion and entertains them accordingly. SJF receives the requests, calculates their size for each cluster of the buildings and allocates resources to them.

3.6. The RR Algorithm

RR is used as the demand scheduling algorithm for cloud environment. It is adapted in this work on the basis of specific time schedules (i.e., time quantum) [30]. The scheduler develops the details of the VMs in an assignment table. In addition, it tries to allocate the jobs equally for the VMs on the DCs; however, it does not confirm completion of their time schedules. Firstly, VM is initiated with the ID of the current VM variable and the requested job is mapped with that VM variable. In some cases, if the IDs of first and last VM resemble then it redefines the current VM ID. It is pre-emptive in nature and if any task has not been completed in the defined time horizon, it reinitiates that task in the next cycle.

Algorithm 1: RR.

Inputs: Set of the incoming requests, Set of the VMs;
Output: *Processing_Time, Request_Time, Response_Time, Cost*;
Initialization: $maxCount = maxVal, VM\_ID = -1$;
for $VM=1$; $VM \leq length(VM\_list)$; $VM++$ **do**
    find (*tasks_assigned* to VM)
    if $VM\_ID == -1$ **then**
        $curr\_Count = 1$;
        else
            $curr\_Count=VM.getCurrentAssign()$;
        end
    end
    endJobCount
    find (*curr_VM_State*)
    state = $VM.getState()$
    if $curr\_Count \leq maxCount$ && *state.equals(AVAILABLE)* **then**
        $maxCount=curr\_Count$
        $VM\_ID = Curr\_VM\_ID$
    end
    return VM_ID
    Calculate *Response_Time*
    Calculate *Request_Time*
    Calculate *Processing_Time*
    Calculate $Cost_{tot}$
end

3.7. The ESCE Algorithm

The ESCE algorithm applies spread spectrum technique and works with the lot of active tasks on the VMs at any time interval [30]. Using this algorithm, the scheduler records the VMs' assignment table while maintaining lists of VM IDs with active jobs on any VM. At each time interval, when the jobs are executed, the VM table is modified. When ESCE starts, active job count is zero; however, on the occurrence of every job, the scheduler determines the VM having the minimum job count. When multiple jobs with minimum count are allocated to the VMs then the first VM is selcted for processing the jobs. Multiple job queues are maintained according to the VMs status. It first allocates the available VMs then it tries to evenly distribute the tasks on them.

Algorithm 2: ESCE.

Inputs: Set of the incoming requests, Set of the VMs;

Output: *Processing_Time, Request_Time, Response_Time, Cost*;

Initialization: *maxCount = maxVal, VM_ID* = 1;

for *VM=1; VM $\leq$ length(VM_list); VM++* **do**

    Task schedular orchestrates the VM assignment table with minimum active job count

    while *task scheduler obtains new incoming jobs* **do**

        Task scheduler checks VM assignment table using active job count for the tasks

        Allocate those jobs to VMs

        Upgrade VM assignment table using active jobs

    end

    if *VM completes allocated jobs* **then**

        upgrade VM assignment table by decrementing the job count

    end

    Return *VM_ID*

    Calculate *Response_Time*

    Calculate *Request_Time*

    Calculate *Processing_Time*

    Calculate *Cost*

end

3.8. The SJF Algorithm

It executes the jobs by considering their shortest size as the priority and that priority is controlled by analyzing consumers' requests' sizes [31]. It assigns the tasks to the VMs based on their regions, fogs and shortest size. This algorithm is preferred over other algorithms because it has minimum latency. The consumers bear less delay and their comfort standards do not compromise more. The requests having the minimum PT or sizes are served first and then VMs are assigned to them respectively. The scheduler allocates the jobs to various VMs and SJF schedules the jobs by allotting the minimum completion time. It also provides higher efficiency and lower turnaround time. In this way, it enhances the system's performance.

Algorithm 3: SJF.

Input: List of the incoming requests, List of the VMs;

Output: *Processing_Time, Request_Time*schedular orchestrates the VM assignment, *Response_Time, Cost*;

Initialization: *maxCount = maxVal, VM_ID* = -1;

for *VM=1; VM $\leq$ length(VM_list); VM++* **do**

    if *length(task$_{i+1}$) $<$ length(task$_i$)* **then**

        Add *task$_{i+1}$* infront of the task queue

    end

    if *length(task$_{VM}$) $== 0$* **then**

        upgrade VM task count

    end

    Return *VM_ID*

    Calculate *Response_Time*

    Calculate *Request_Time*

    Calculate *Processing_Time*

    Calculate *Cost$_{tot}$*

end

---

4. Simulation Results

In this section, six regions are considered for conducting the simulations of the proposed system. Each region is comprised of a cluster of buildings varying between 50 to 150 and every building has 80 to 100 homes in it. The number of buildings and homes are specified as random for validating the system performance. Simulations are performed for 24 h in a complete day by considering the performance parameters: RT, PT, RPH and cost (i.e., cost of VM, total number of MG installed and total data transfer). These parameters are computed through the Equations (1)–(7). The information about regions is described in Table 2.

Table 2. Region information.

Region	Region Id	Number of Users
North America	0	80 M
South America	1	20 M
Europe	2	60 M
Asia	3	27 M
Africa	4	5 M
Ocenia	5	8 M

For performing the simulations, optimized RT resource allocation policy is used whereas the load balancing algorithms are used as RR, ESCE and our proposed algorithm SJF. Our proposed algorithm gives the minimum latency as compared to the other two algorithms. It helps in maintaining consumers' comfort preferences. As there exists randomness for multiple epochs of these algorithms, that is why, we have taken the average of 10 times. The simulations of the proposed work are based on the following scenarios.

4.1. Scenario 1: Resource Allocation Using MGs

Scenario 1 is comprised of the MGs without any storage resources (i.e., batteries) whereas scenario 2 is comprised of storage resources along with the MGs. Each scenario is further incorporating the two sub-scenarios: 25 VMs with single DC and 50 VMs with two DCs. Now, the RT of all algorithms is computed first then all other performance metrices are calculated later.

4.1.1. Scenario 1(1): 25 VMs for Single DC

In scenario *1(1)*, 25 VMs with one DC are considered for intelligent resource utilization. Among the six clusters, the RT of all fogs is optimized using the MGs installed in the regions. It is observed that during peak hours, users have higher demands as compared to low peak hours. The details of peak and off peak hours regarding each region are taken from [30]. Since the fog accomodates efficiency in latency while communicating to the cloud environment, this framework gives a reliable solution in the energy management and resource allocation as displayed in Figure 2. The RT of the SJF algorithm is more efficient than the RR and ESCE. SJF processes the requests with the minimum completion time and entertains the other requests which are having the largest PT. Further, RR entertains all the cloudlets in a cyclic order based on their request time. In addition, ESCE checks the free VMs and equally distributes all requests to them. Two MGs are installed in each region to fulfill the demands of consumers. The RT obtained by the ESCE, RR and SJF algorithms throughout the simulations is: 65 ms, 58 ms and 57 ms for fog 1; 60 ms, 58 ms and 57 ms for fog 2; 64 ms, 58 ms and 57 ms for fog 3; 63 ms, 58 ms and 57 ms for fog 4; 53 ms, 50 ms and 49 ms for fog 5; and 58 ms, 55 ms and 54 ms for fog 6. The RT is totally based on the number of the requests for each hour. So, the requests are based on the population of all clusters as shown in Table 2. For fog 1, SJF is 13% more efficient than the ESCE; whereas it is 2% more efficient than RR. For fog 2, it is 5% and 2% more intelligent than the previous two algorithms. SJF outperforms ESCE and RR up to 11% and 2% for fog 3. In fog 4, SJF beats the previous algorithms: ESCE and RR up to 10% and 2%. For fog 5, SJF obtains the efficient RT of 8% and

2% as compared to ESCE and RR. SJF is 7% and 2% more efficient than ESCE and RR algorithm for fog 6. It is observed that SJF can handle more requests easily and outperforms the previous algorithms as shown in Figure 4 because the regions: 0, 2 and 3 have the maximum population and it shows better results for them as compared to the previous algorithms.

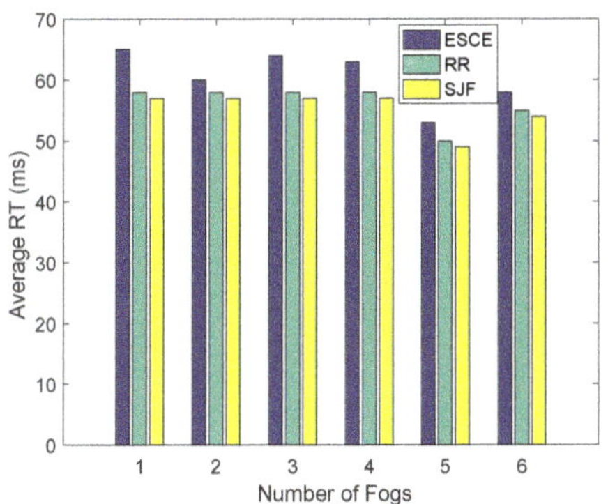

Figure 4. Average Response Time (RT) of All Fogs in Scenario *1(1)*.

The average PT of each fog according to the users' requests is displayed in Figure 5 for the whole day using the aforementioned algorithms. The average PT of the fog 1 is 13 ms, 9 ms and 8 ms using the ESCE, RR and SJF algorithms. Fog 2 also has the effective PTs which are optimized by these algorithms as 14 ms, 9 ms and 8 ms using ESCE, RR and SJF. SJF outperforms the other algorithms for computing the requests of region 3 consumers in 13.5 ms, 8 ms and 9 ms using the above-mentioned algorithms. ESCE, RR and SJF obtain the optimized PT from the fog 4 up to 13.8 ms, 9 ms and 8 ms. Fog 5 obtains the efficient PT using the SJF as compared to the ESCE and RR upto 7 ms, 10 ms and 8 ms. All algorithms provide 5 ms, 4 ms and 3 ms optimized PT for Fog 6. From these results, it can be concluded that SJF gives more optimized results for all of the fogs in the respective regions. All regions have sufficient population and the proposed algorithm gives efficient results as compared to the previous algorithms. For example: it outperforms the ESCE and RR up to 39% and 2% for fog 1; 37% and 2% for fog 2; 41% and 2% for fog 3; 43% and 2% for fog 4; 93% and 13% for fog 5; and 40% and 25% for fog 6.

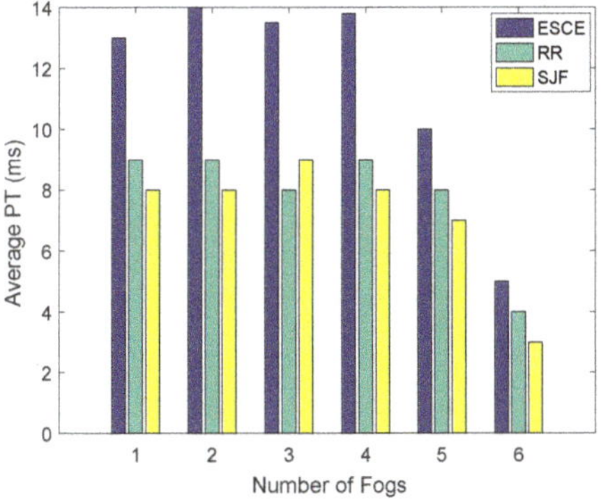

Figure 5. Average Processing Time (PT) of All Fogs in Scenario *1(1)*.

Consumers' RPH is shown in Figure 6 for all the fogs installed in the respective regions. Fog loading is done as per the number of requests received from the buildings. The number of buildings and homes are considered random; however, their requests are taken as fixed in each time interval. The fog loading phase is initiated after the requests' receiving phase and then tasks' computation is performed. The maximum population is considered from each region for getting the maximum number of requests and the number of VMs, MGs and other relevant services are available on the fog and cloud DCs for fulfilling the requests. The requests from each region are based on the number of consumers living in that region as shown in Figure 6.

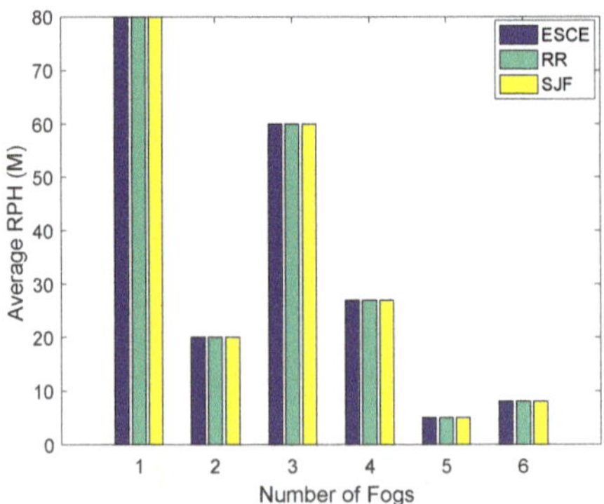

Figure 6. Request Per Hour (RPH) of the Consumers in Scenario *1(1)*.

Using the six fogs for six regions, the proposed system's total cost is comprised of the VMs cost, MGs cost and data transfer cost. The number of MGs, VMs and all the other allocations relevant to users' requests along with their costs are calculated and analyzed. There is a trade-off in the VMs cost using the SJF algorithm; since it receives the tasks with the small size priority and allocates those tasks to the available VMs, it has a high cost. The VMs cost obtained by the ESCE, RR and SJF is 300\$, 478\$ and 300\$. MG cost is computed as: 2731.94\$, 2400\$ and 2734.31\$ respectively. The data transfer cost is calculated as 5795.75\$, 3800.53\$ and 5795.79\$. In addition, computation cost for the MG and data transfer cost for the other two algorithms is not very high because they do not consider the request's size priority. Aggregated cost for all three algorithms is shown in Figure 7.

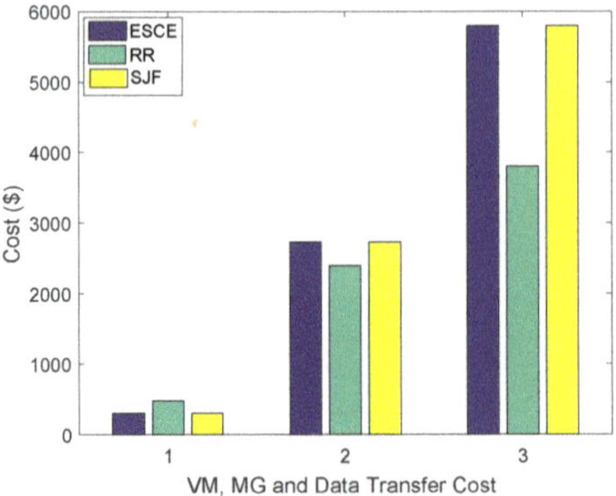

Figure 7. VM, MG and Data Transfer Cost.

4.1.2. Scenario 1(2): 50 VMs with Two DCs

In this scenario, 50 VMs are used with two DCs in order to check the efficiency of the proposed C2F2C framework. First the RT of all the fogs is computed as shown in Figure 8. The RT of all fogs calculated according to the above-mentioned algorithms: ESCE, RR and SJF is 32.5 ms, 29 ms and 28.25 ms for fog 1; 30 ms, 29 ms and 28.25 ms for fog 2; 32 ms, 29 ms and 28.25 ms for fog 3; 31.5 ms, 29 ms and 28 ms for fog 4; 26.5 ms, 24 ms and 23 ms for fog 5; and 24 ms, 23 ms and 21 ms for fog 6.

In this case, SJF outperforms all algorithms efficiently as it is also observed from all of the previous fogs results. SJF outperforms ESCE and RR: 14% and 3% for fog 1; 6% and 3% for fog 2; 12% and 3% for fog 3; 12% and 3% for fog 4 similar to fog 3; 14% and 3% for fog 5; and 3% using both algorithms for fog 6, respectively.

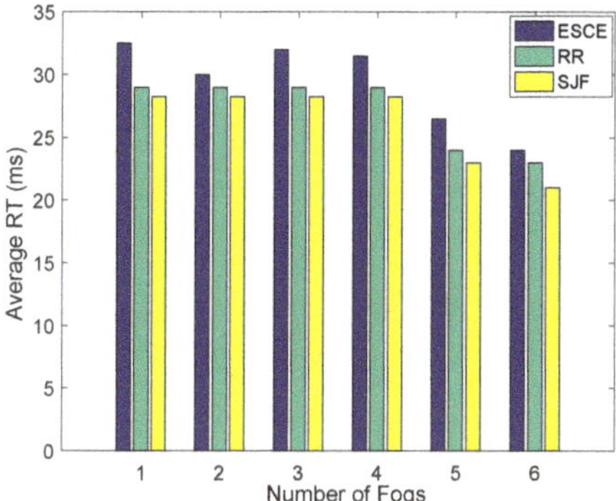

Figure 8. RT of All the Fogs in Scenario *1(2)*.

After computing the RT of the scenario, the PT of each fog is also computed as displayed in Figure 9. The PT for fog 1 is computed as: 7.5 ms, 4.5 ms and 4 ms using these algorithms. For fog 2–6, it is calculated as: 7 ms, 4.5 ms and 4 ms for fog 2; 6.55 ms, 4.5 ms and 4 ms for fog 3; 6.54 ms, 4.5 ms and 4 ms for fog 4; 5 ms, 4.5 ms and 4 ms for fog 5; and 2.5 ms, 2 ms; and 1.5 ms for fog 6 through ESCE, RR and SJF algorithms. These algorithms have optimized the proposed work efficiently, especially SJF algorithm performs better then the other two algorithms. For fog 1–6, it performs 47% and 12%; 43% and 12%; 39% and 12%; 82% and 12%; and 94% and 25% better than ESCE and RR. This scenario gives the more optimal results in terms of the resource allocation because more resources are utilized and consumers are entertained more efficiently as compared to the first scenario.

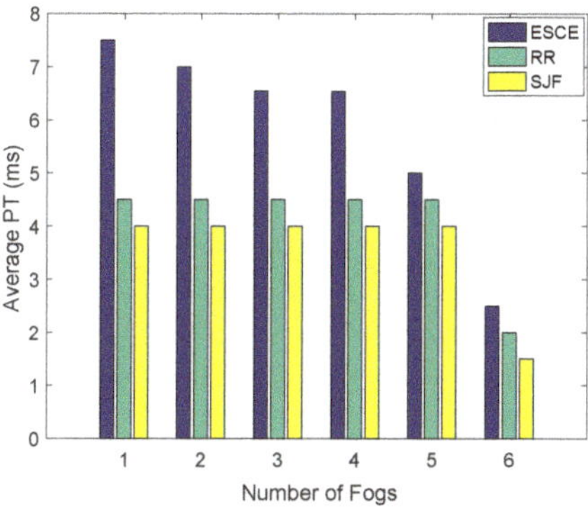

Figure 9. PT of All the Fogs in Scenario *1(2)*.

Figure 10 shows the RPH of the consumers which is kept the same as in scenario 1. For the sake of simplicity, RPH is kept similar to all scenarios in this work.

As mentioned above, the aggregated cost is comprised of the VMs cost, MGs cost and data transfer cost in this system. For this scenario, the resources are used twice as compared to the resources used in scenario *1(2)*, so, it costs almost double to the scenario *1(1)*; however, it maximizes the RT and PT of the consumers' requests. The total cost for this scenario is shown in Figure 11. The cost computed for VMs, MGs and data transfer is: 600$, 756$ and 600$; 5463.88$, 4800$ and 5468.62$; and 11591.5$, 7601.06$

and 11591.58\$. This scenario has achieved more optimized results as compared to the scenario *1(1)*; however, it compromises the cost.

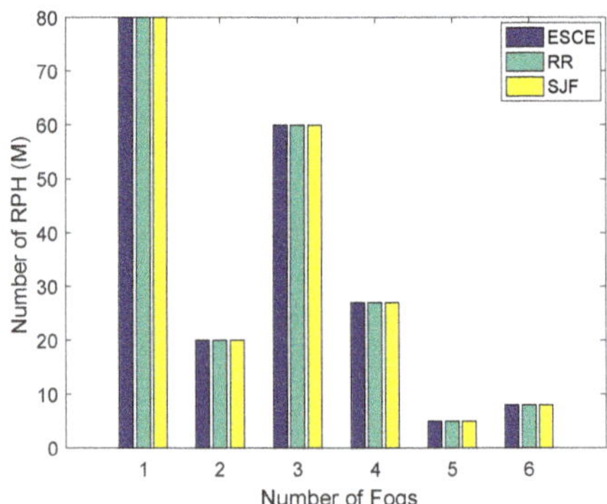

Figure 10. RPH of All the Consumers in Scenario *1(2)*.

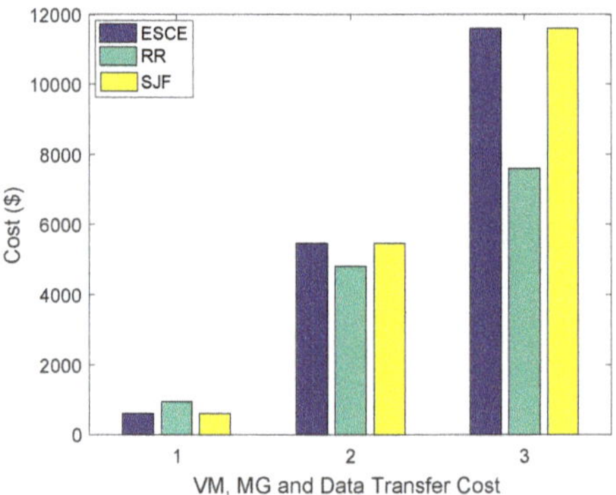

Figure 11. Virtual Machines (VM), Microgrids (MG) and Data Transfer Cost.

4.2. Scenario 2: Resource Allocation Using MGs and Power Storage Devices

The second use case scenario for the proposed framework is demonstrated in Figure 12. In this scenario, a power storage system (i.e., a battery system) is integrated for improving the RT, PT and cost of the consumers' requests. The components of the system are smart homes, power storage devices, smart solar panel, wind power, electric vehicle and retail super store. The smart homes' users request the electricity from the fog at the regional level where fog is installed to fulfill their demands. Fog is connected to the MGs, electric vehicles and battery storage resources which are used to facilitate the consumers' requests. Fog first sends requests to MGs or battery storage devices to fulfill the demands of the consumers. If these are sufficient then it does not take any power from the utility. In the case that there is surplus power, then it is stored in the batteries and it can be sold back to the utility for the sake of the power trading. The stored power is also utilized in the peak hours in this work. Power trading is not the focus of this work. Here, the focus of this work is making the power system more sustainable and flexible for residential buildings. Electric vehicle and retail super store are directly connected to the fog and are solely based on the MGs and battery storage devices for procuring electricity services. Because of this, these are considered as sustainable stations in any region. Electric vehicles are used for fulfilling the travel requirements of the consumers in any residential region. These are not directly connected to the main grid for utilizing the power from main grid. These are consumers and prosumers of the residential area. The parameters are considered as the same as those considered in scenario *1(1)* and *1(2)*.

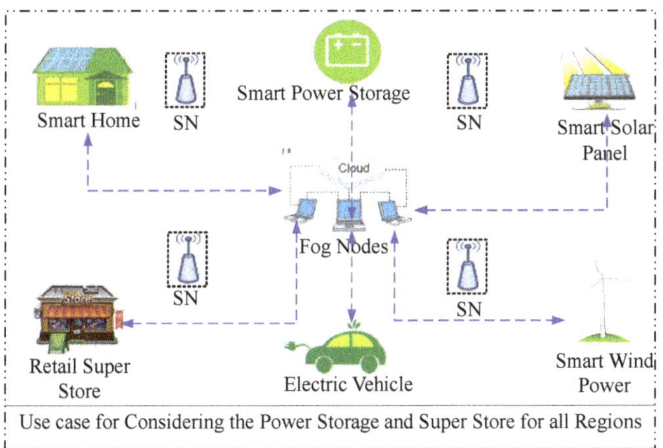

Figure 12. Working of Scenario 2 in Residential Buildings using Battery Storage.

4.2.1. Scenario 2(1): 25 VMs for Single DC and Power Storage

In this scenario, the effect of battery storage devices has been verified through simulations using the same performance metrices. Firstly, RT of the system is computed which is shown in Figure 13. RT of the system is optimized using the ESCE, RR and SJF algorithms which are used in the previous scenario. The RT of fog 1 is 50 ms, 43 ms and 42 ms; fog 2 is 45 ms, 43 ms and 42 ms; fog 3 is 49 ms, 43 ms and 42 ms; fog 4 is 48 ms, 43 ms and 42 ms; fog 5 is 38 ms, 35 ms and 34 ms; and fog 6 is 43 ms, 40 ms and 39 ms, respectively. Here SJF beats the other algorithms: ESCE and RR up to 16% and 14% for fog 1; 7% and 14% for fog 2; 15% and 14% for fog 3; 13% and 14% for fog 4; 11% and 14% for fog 5; and 10% and 14% for the fog 6, respectively.

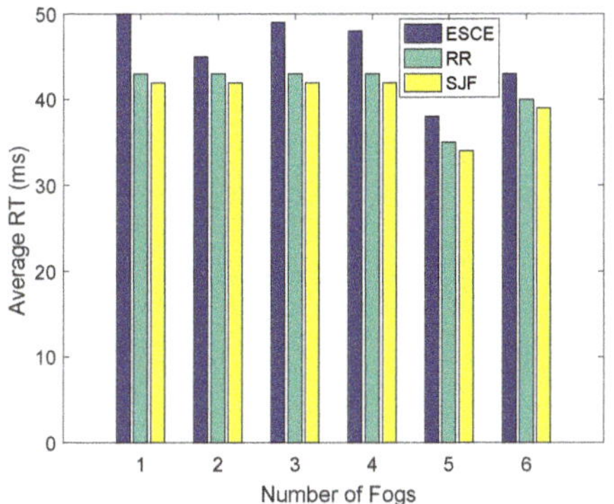

Figure 13. RP of All the Consumers in Scenario *2(1)*.

Analysing the RT of the scenario, it is observed that SJF outperforms the other two algorithms due to the incorporation of the battery storage resources. RPH of this scenario is displayed in Figure 14 which is kept as the same according to the population of each region.

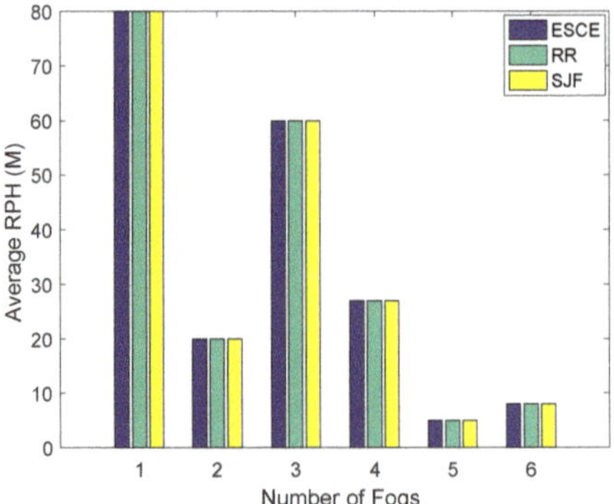

Figure 14. RPH of All the Consumers in Scenario *2(1)*.

Now the PT of each fog has been optimized for this scenario using these algorithms: ESCE, RR and SJF as shown in Figure 15. Each fog has optimal PT as: 11.5 ms, 7.5 ms and 6.5 ms for fog 1; 12.5 ms, 7.5 ms and 6.5 ms for fog 2; 11 ms, 7.5 ms ad 6.5 ms for fog 3; 11.3 ms, 7.5 ms and 6.5 ms for fog 4; 8.5 ms, 7.5 ms and 6.5 ms for fog 5; and 3.5 ms, 2.5 ms and 2.25 ms for fog 6. SJF beats the ESCE and RR upto 44% and 14% for fog 1; 46% and 14% for fog 2; 41% and 14% for fog 3; 43% and 14% for fog 4; 24% and 14% for fog 5; and 36% and 10% for fog 6.

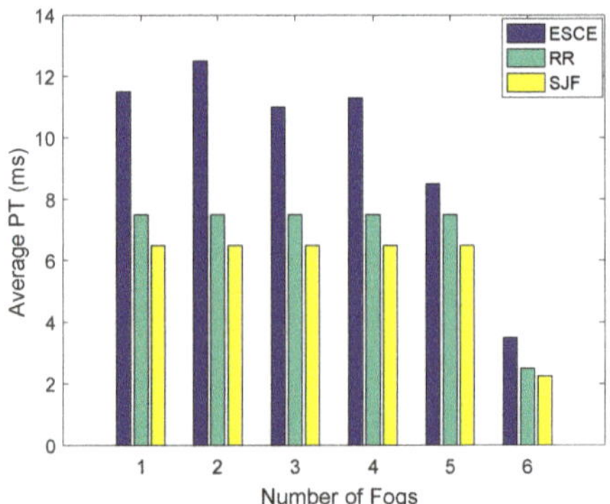

Figure 15. PT of All the Consumers in Scenario *2(1)*.

The cost is also minimized up to 15% of the total cost of the scenario *1(1)* with the incorporation of the battery storage resources. It is displayed in Figure 16. Overall, RT and PT of this scenario have improved up to 15% with the integration of the battery storage resources as compared to scenario *1(1)*.

4.2.2. Scenario 2(2): 50 VMs for Single DC and Battery Storage

In this scenario, battery storage devices are used with the increased number of VMs and DCs as compared to the scenario *2(1)*. The purpose of this scenario is also based on the optimization for the similar performance parameters. The RT is computed as: 30.5 ms, 27.5 ms and 26.75 ms for fog 1; 28.5 ms, 27.5 ms and 26.75 ms for fog 2; 30.5 ms, 27.5 ms and 26.75 ms for fog 3; 30 ms, 27.5 ms and 26.75 ms for fog 4; 25 ms, 22.5 ms and 21.5 for fog 5; and 22.5 ms, 21.5 ms and 19.5 ms for fog 6 as demonstrated in Figure 17. From these three algorithms, our proposed algorithm performs the best. It outperforms ESCE and RR up to: 13% and 3% for fog 1; 7% and 3% for fog 2; 13% and 3% for fog 3 which is similar to fog 1; 11% and 3% for fog 4; 14% and 3% for fog 5; and 14% and 10% for fog 6.

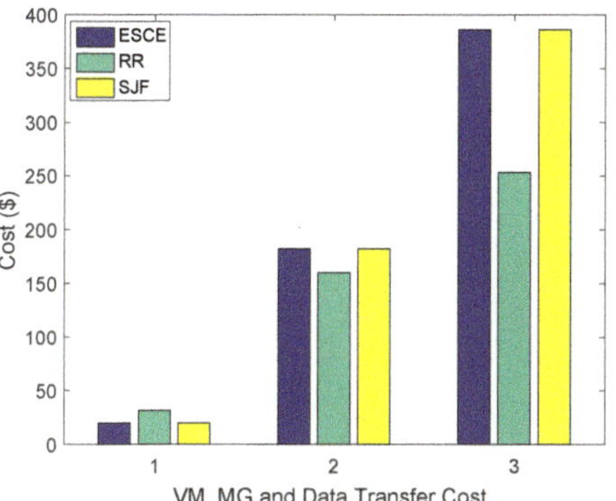

Figure 16. VM, MG and Data Transfer Cost.

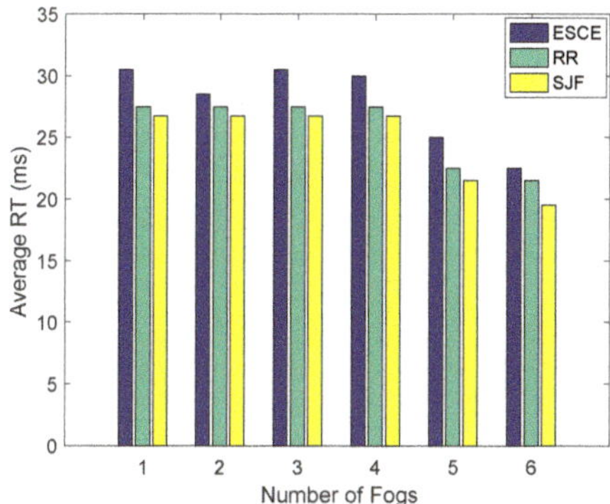

Figure 17. RP of All the Consumers Scenario *2(2)*.

RPH is also kept similar in this case as displayed in Figure 18 for each region throughout the simulation.

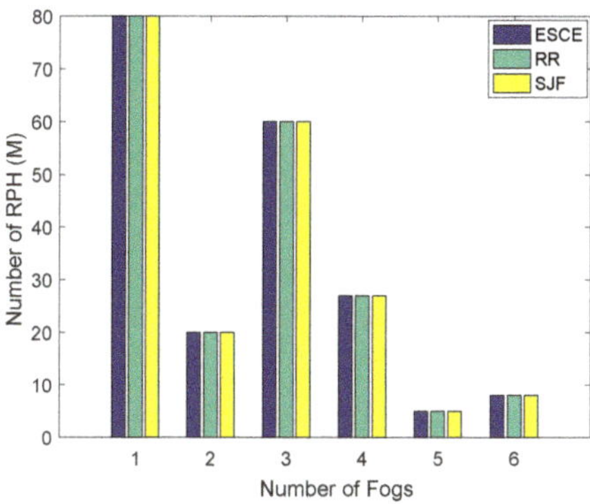

Figure 18. RPH of All the Consumers in Scenario *2(2)*.

After discussing the RPH and RT, PT of the proposed framework is computed which is shown in Figure 19. The PT for all fogs is computed as: 6 ms, 3 ms, and 2.5 ms for fog 1; 5.5 ms, 3 ms and 2.5 ms for fog 2; 5.15 ms, 3 ms, and 2.5 ms for fog 3; 5.15 ms, 3 ms and 2.5 ms for fog 4; 3.5 ms, 3 ms and 2.5 ms for fog 5; and 1.5 ms, 1 ms and 0.75 ms for fog 6. In this case, previous algorithms compromise the performance where SJF performs efficiently using the storage devices. SJF outperforms ESCE and RR up to: 59% and 50% for fog 1; 55% and 50% for fog 2; 52% and 50% for fog 3; 52% and 50% for fog 4 which is similar to fog 3; 29% and 3% for fog 5; and 50% and 25% for fog 6, respectively.

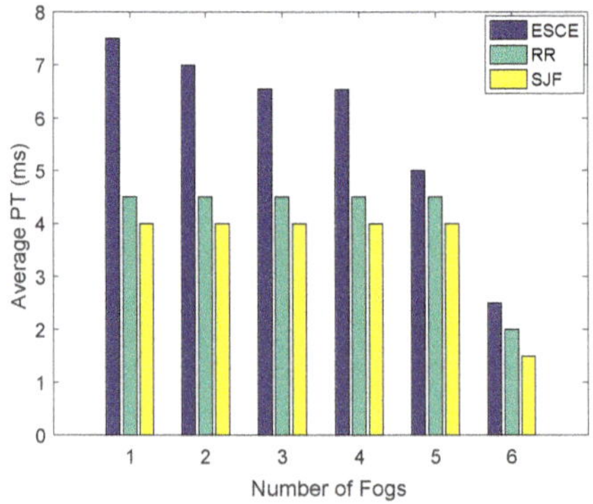

Figure 19. PT of All the Consumers in Scenario *2(2)*.

The total system cost is shown in Figure 20. It improves the cost up to 15% of the scenario *1(2)* by installing the battery storage resources and by considering the electric vehicles.

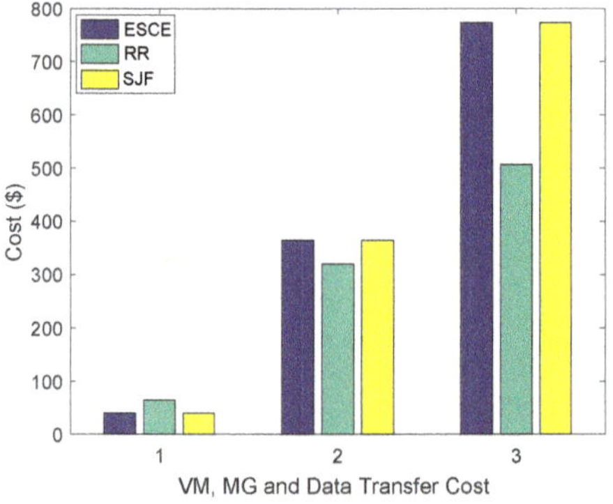

Figure 20. VM, MG and Data Transfer Cost.

4.3. Critical Analysis

After conducting the simulations of the proposed framework, the following pros and cons are analysed from the fog and cloud integrated environment, scheduling algorithms and their performance parameters' effects.

- The idea of fog is used for minimizing the latency; however, it has minimum resources which requires more alternatives.
- After conducting the simulations of the proposed algorithm, it is observed that it minimizes the latency. On the other hand, it utilizes more resources which may be a burden to the whole system.
- Using the scenarios *1(2)* and *2(2)*, although more of the system's resources increase the PT and RT of the system, they also increase the overall system cost.
- The RT and PT of the system increase 50% more efficient performance in scenarios *1(2)* and *2(2)* as compared to the performance of scenario I; however, they compromise the MG and data transfer cost.

5. Conclusions, Future Work and Future Challenges

A C2F2C based framework has been proposed and implemented for intelligently allocating the resources in the residential buildings. This framework is based on three layers where consumers' requests have been considered constant for every hour of a day. Two control parameters: clusters of buildings and load requests are considered from any region. Whereas the performance parameters are RPH, RT, PT and the cost (i.e., VMs, MGs and data transfer). The optimization of these parameters have been performed by the SJF, ESCE and RR algorithms in this study. Two scenarios are considered for simulations: resource allocation using MGs and resource allocation using MGs, electric vehicles and power storage resources. Each scenario is further categorized into two other scenarios: one DC with 25 VMs and two DCs with 50 VMs. The performance of the proposed framework has increased, i.e., 50% in scenarios *1(2)* and *2(2)* as compared to the scenarios *1(1)* and *2(1)* in terms of RT and PT. Tradeoff occurs in the PT, as our system has processed and received more requests and it takes a lot of time to process the consumers' demands. In scenarios *1(2)* and *2(2)*, tradeoff occurs in cost due to maximum resource utilization using proposed algorithm. Simulation results also show that our technique has outperformed the prior techniques in terms of the aforementioned parameters as shown in the simulation section.

We described a C2F2C framework with simulations to support our idea. We believe that there are many other modern solutions; e.g., use of SDNs or blockchain that could replace the role of the cloud. This is one of our future interests.

For substantiating the implications of this study, the following future challenges need to be tackled intelligently.

- How can individual appliances be made smarter in order to maintain the comfort level of consumers in any region using the proposed framework?
- How can the PT be improved by increasing the number of requests?
- How will cost be optimized using the multiple resources as considered in scenarios *1(2)* and *2(2)*?
- Which strategies will be applicable for optimizing the tradeoffs of this work, i.e., tradeoffs between PT, RT and cost?
- What will be the effects of multi-objective optimization for mitigating the existing work's tradeoffs?
- What will be the effects of this framework on energy management if it will be investigated with the big data analytics along with other techniques?

Author Contributions: S.J. and N.J. proposed and implemented the main idea. T.S. and Z.W. performed the mathematical modeling and wrote the simulation section. A.R. and A.H. organized and refined the manuscript.

Funding: This research received no external funding.

Acknowledgments: This work was supported by AI and Data Analytics (AIDA) Lab Prince Sultan University Riyadh Saudi Arabia.

Conflicts of Interest: All authors are agreed on this work. The authors declare no conflict of interest.

References

1. Gungor, V.C.; Sahin, D.; Kocak, T.; Ergut, S.; Buccella, C.; Cecati, C.; Hancke, G.P. Smart grid technologies: Communication technologies and standards. *IEEE Trans. Ind. Inform.* **2011**, *7*, 529–539. [CrossRef]
2. Opower. Available online: http://www.opower.com/ (accessed on 18 March 2018).
3. C3 Energy. Available online: https://c3energy.com/ (accessed on 18 March 2018).
4. Samadi, P.; Mohsenian-Rad, H.; Schober, R.; Wong, V.W. Advanced demand side management for the future smart grid using mechanism design. *IEEE Trans. Smart Grid* **2012**, *3*, 1170–1180. [CrossRef]
5. Fang, X.; Misra, S.; Xue, G.; Yang, D. Smart grid-The new and improved power grid: A survey. *IEEE Commun. Surv. Tutor.* **2012**, *14*, 944–980. [CrossRef]

6.   Zhao, J.; Wan, C.; Xu, Z.; Wang, J. Risk-based day-ahead scheduling of electric vehicle aggregator using information gap decision theory. *IEEE Trans. Smart Grid* **2017**, *8*, 1609–1618. [CrossRef]

7.   Vaquero, L.M.; Rodero-Merino, L. Finding your way in the fog: Towards a comprehensive definition of fog computing. *ACM SIGCOMM Comput. Commun. Rev.* **2014**, *5*, 27–32. [CrossRef]

8.   Kobo, H.I.; Abu-Mahfouz, A.M.; Hancke, G.P. A Survey on Software-Defined Wireless Sensor Networks: Challenges and Design Requirements. *IEEE Access* **2017**, *5*, 1872–1899. [CrossRef]

9.   Chiang, M.; Zhang, T. Fog and IoT: An overview of research opportunities. *IEEE Internet Things J.* **2016**, *3*, 854–864. [CrossRef]

10.   Luo, F.; Zhao, J.; Dong, Z.Y.; Chen, Y.; Xu, Y.; Zhang, X.; Wong, K.P. Cloud-based information infrastructure for next-generation power grid: Conception, architecture, and applications. *IEEE Trans. Smart Grid* **2016**, *7*, 1896–1912. [CrossRef]

11.   Xing, H.; Fu, M.; Lin, Z.; Mou, Y. Decentralized optimal scheduling for charging and discharging of plug-in electric vehicles in smart grids. *IEEE Trans. Power Syst.* **2016**, *31*, 4118–4127. [CrossRef]

12.   Gan, L.; Topcu, U.; Low, S.H. Optimal decentralized protocol for electric vehicle charging. *IEEE Trans. Power Syst.* **2013**, *28*, 940–951. [CrossRef]

13.   Okay, F.Y.; Ozdemir, S. A fog computing based smart grid model. In Proceedings of the 2016 International Symposium on Networks, Computers and Communications (ISNCC), Hammamet, Tunisia, 11–13 May 2016; pp. 1–6.

14.   Javaid, S.; Javaid, N.; Tayyaba, S.; Sattar, N.A.; Ruqia, B.; Zahid, M. Resource allocation using Fog-2-Cloud based environment for smart buildings. In Proceedings of the 14th IEEE International Wireless Communications and Mobile Computing Conference (IWCMC-2018), Limassol, Cyprus, 25–29 June 2018; pp. 1–6.

15.   Chekired, D.A.; Khoukhi, L. Smart Grid Solution for Charging and Discharging Services Based on Cloud Computing Scheduling. *IEEE Trans. Ind. Inform.* **2017**, *13*, 3312–3321. [CrossRef]

16.   Li, Y.; Chen, M.; Dai, W.; Qiu, M. Energy optimization with dynamic task scheduling mobile cloud computing. *IEEE Syst. J.* **2017**, *11*, 96–105. [CrossRef]

17.   Moghaddam, M.H.Y.; Leon-Garcia, A.; Moghaddassian, M. On the performance of distributed and cloud-based demand response in smart grid. *IEEE Trans. Smart Grid* **2017**, *9*, 5403–5417.

18.   Cao, Z.; Lin, J.; Wan, C.; Song, Y.; Zhang, Y.; Wang, X. Optimal cloud computing resource allocation for demand side management in smart grid. *IEEE Trans. Smart Grid* **2017**, *8*, 1943–1955.

19.   Kumar, N.; Vasilakos, A.V.; Rodrigues, J.J. A multi-tenant cloud-based DC nano grid for self-sustained smart buildings in smart cities. *IEEE Commun. Mag.* **2017**, *55*, 14–21. [CrossRef]

20.   Al Faruque, M.A.; Vatanparvar, K. Energy management-as-a-service over fog computing platform. *IEEE Internet Things J.* **2016**, *3*, 161–169. [CrossRef]

21.   Yaghmaee, M.H.; Moghaddassian, M.; Leon-Garcia, A. Autonomous two-tier cloud-based demand side management approach with microgrid. *IEEE Trans. Ind. Inform.* **2017**, *13*, 1109–1120. [CrossRef]

22.   Sheikhi, A.; Rayati, M.; Bahrami, S.; Ranjbar, A.M. Integrated demand side management game in smart energy hubs. *IEEE Trans. Smart Grid* **2015**, *6*, 675–683. [CrossRef]

23.   Chekired, D.A.; Khoukhi, L.; Mouftah, H.T. Decentralized Cloud-SDN Architecture in Smart Grid: A Dynamic Pricing Model. *IEEE Trans. Ind. Inform.* **2017**, *14*, 1220–1231. [CrossRef]

24.   Reka, S.S.; Ramesh, V. Demand side management scheme in smart grid with cloud computing approach using stochastic dynamic programming. *Perspect. Sci.* **2016**, *8*, 169–171. [CrossRef]

25.   Fang, B.; Yin, X.; Tan, Y.; Li, C.; Gao, Y.; Cao, Y.; Li, J. The contributions of cloud technologies to smart grid. *Renew. Sustain. Energy Rev.* **2016**, *59*, 1326–1331. [CrossRef]

26.   Kumar, N.; Zeadally, S.; Misra, S.C. Mobile cloud networking for efficient energy management in smart grid cyber-physical systems. *IEEE Wirel. Commun.* **2016**, *23*, 100–108. [CrossRef]

27.   Yao, J.; Liu, X.; Zhang, C. Predictive electricity cost minimization through energy buffering in data centers. *IEEE Trans. Smart Grid* **2014**, *5*, 230–238. [CrossRef]

28.   Mohamed, N.; Al-Jaroodi, J.; Jawhar, I. Service-Oriented Big Data Analytics for Improving Buildings Energy Management in Smart Cities. In Proceedings of the 2018 14th International Wireless Communications & Mobile Computing Conference (IWCMC), Limassol, Cyprus, 25–29 June 2018; pp. 1243–1248.

29. Devi, D.C.; Uthariaraj, V.R. Load balancing in cloud computing environment using improved weighted round robin algorithm for nonpreemptive dependent tasks. *Sci. World J.* **2016**, *2016*, 3896065. [CrossRef] [PubMed]
30. Wickremasinghe, B.; Buyya, R. Cloud Analyst: A CloudSim-based tool for modeling and analysis of large scale cloud computing environments. *MEDC Proj. Rep.* **2009**, *22*, 433–659.
31. Kaur, R.; Kinger, S. Analysis of job scheduling algorithms in cloud computing. *Int. J. Comput. Trends Technol.* **2014**, *9*, 379–386. [CrossRef]

Article

Energy Retrofitting Effects on the Energy Flexibility of Dwellings

Francesco Mancini [1],* and Benedetto Nastasi [2]

[1] Department of Planning, Design and Technology of Architecture, Sapienza University of Rome, Via Flaminia 72, 00196 Rome, Italy
[2] Department of Architectural Engineering & Technology, TU Delft University of Technology, Julianalaan 134, 2628BL Delft, The Netherlands
* Correspondence: francesco.mancini@uniroma1.it; Tel.: +39-06-4991-9172

Received: 26 June 2019; Accepted: 18 July 2019; Published: 19 July 2019

Abstract: Electrification of the built environment is foreseen as a main driver for energy transition for more effective, electric renewable capacity firming. Direct and on-time use of electricity is the best way to integrate them, but the current energy demand of residential building stock is often mainly fuel-based. Switching from fuel to electric-driven heating systems could play a key role. Yet, it implies modifications in the building stock due to the change in the temperature of the supplied heat by new heat pumps compared to existing boilers and in power demand to the electricity meter. Conventional energy retrofitting scenarios are usually evaluated in terms of cost-effective energy saving, while the effects on the electrification and flexibility are neglected. In this paper, the improvement of the building envelope and the installations of electric-driven space heating and domestic hot water production systems is analyzed for 419 dwellings. The dwellings database was built by means of a survey among the students attending the Faculty of Architecture at Sapienza University of Rome. A set of key performance indicators were selected for energy and environmental performance. The changes in the energy flexibility led to the viable participation of all the dwellings to a demand response programme.

Keywords: energy flexibility; retrofitting interventions; residential consumption; demand response; electrification in the built environment

---

1. Introduction

Stagnation in energy efficiency improvements and up and down trends of carbon emissions call for ambitious targets. This is why EU sets energy efficiency at least 32.5% by 2030 [1]. Buildings are responsible for 41.7% of energy consumption [2]. Therefore, larger renovation rates, renewable heating, and related reduced pollutants [3] are actions enhanced by automation and smart controls in [4]. At present, the increase of open data availability and its inclusion in countries policies for renewable energy assessment [5], and to improve the energy performance of the building stock and data collecting of operational phase [6].

On the side of renewable energy, the European Commission set that the contribution of renewables in 2030 will cover 32% of final energy consumptions. Today, the major part of the energy system is based on fossil fuels. By mid-century, this will change radically with the large-scale electrification of the energy system driven by the deployment of renewables and fully developed alternative fuel options [7]. Regarding the electric sector, this ambitious goal implies that more than half of the European electricity demand will be met by renewables. Large RES integration and control strategies are crucial for actual decarbonization at the national [8] as well as urban scale [9]. It entails the deep testing and calibration of multi-energy flows modelling [10] to manage stochastic behavior of renewables. Together with this modelling part, actual involvement of consumers and producers is done by Smart Grid approach [11] and Demand Response (DR) Program [12].

DR aims at adjusting consumers' demand to the energy flow or market price thanks to Demand Side Management (DSM) by means of incentive paid to them [13]. Therefore, buildings connected to the grid can be players by offering peak shaving or balancing services, and their value is weighted on the amount of flexible power for each single user. It implies in building such a mechanism, the crucial role played by probabilistic modelling is beyond the design of the load profile for buildings [14] and beyond the incentive schemes for the market [15]. Dwellings have a small flexible power when considered alone, while a group of them offers the chance to design several pathways of participation to the market, from the local installation of storage [16] to tariff definition [17], from their aggregation as robust and optimized equivalent load [18] up to the neighborhood scale [19], and the sum of their equipment as well [20].

When heating systems are electric-driven, a great potential of flexibility occurs and is enhanced by accounting for building thermal inertia [21] or installation of further storage means. The share of electric-driven heating and cooling on the total electricity and whole energy demand is, therefore, linked to the climatic conditions of building location, its characteristics, and the occupants' behavior [22]. As a matter of fact, the intended use of the building is where electric DHW production is installed and number of occupants are the main parameters affecting the final load value [23]. Moreover, those parameters affect the shiftable loads as well and can be generalized to all the electric appliances and devices used in dwellings [24].

With the aim to assess the potential of flexibility in dwellings, the temporal changes in the residential sector and the trends of energy efficiency policies is a key player [25]. In [26], determinants and trends of energy consumption in EU dwellings is analyzed accounting for impact and effectiveness of energy efficiency policies already implemented. A gap in the analysis is the effect of retrofitting strategies on the current and future flexibility potential, especially when the fuel switching is foreseen and, subsequently, a massive electrification is promoted.

The same gap is identified in the recent literature for Key Performance Indicators (KPI) elaboration. They were generally built to check viability of refurbishment considering economic output such as Wang et al. [27] on a life cycle base or in [28] where net present value, the payback period, and energy savings are taken as the main performance indicators of the retrofitting plan. Wu et al. [29] extended the life cycle analysis to GHG emissions. while Penna et al. [30] checked the thermal comfort ensured by the new building scenarios. Asadi et al. [31] provided an optimization process for environmental and economic performances and Guardigli et al. [32] assessed the economic sustainability of various project alternatives with net present value and global cost, but including social aspects as well. Therefore, energy saving and economic trade-off are largely used in literature [33], especially if the building stock owner is a Public Administration [34] or Social Housing corporate [35]. Only deep renovations seem to be the place for further research questions. However, they are intended to provide more detailed economic outcomes such as in [36], where Niemelä et al. checked cost-optimal retrofit measures in typical Finnish buildings or in [37], where Ortiz et al. designed cost-optimal scenarios for retrofitting residential buildings in Barcelona on global cost evaluation for building lifespan. Assessing the effects of energy retrofitting on flexibility is still missing and is investigated by the authors of the current paper by means of dedicated KPIs.

Four indicators are built: (i) the energy consumption; (ii) renewable energy use; (iii) local carbon emissions; and (iv) flexible loads amount. In detail, local emissions were considered instead of global emissions since there is a clear correlation between those latter ones and the energy consumption due to the calculation methods, while their allocation is not specified [38].

The presence of conventional KPIs, i.e., the first three ones linked to a fourth, the new one, is useful to see the eventual correlation or dependence on each other. Indeed, the flexible loads amount and the observation of its link with the other indicators is the novel contribution of this study. That metric is neglected for dwellings, being the gap to be filled by this study. To summarize, the present study analyzes the effects of different energy retrofitting solutions on the flexibility potential of dwellings. A sample of 419 dwellings in Italy is built thanks to a survey among the students of the Faculty of

Architecture at Sapienza University of Rome. A questionnaire designed for non-energy experts is used to collect the data. Then, simulation scenarios provide the outcome of the new energy demand, its new proportion among fuel and electricity based, and the updated share between the aforementioned different flexible loads.

2. Materials and Methods

In this paper, the data collection to investigate the energy consumptions in residential buildings was undertaken by means of a questionnaire survey. Data collection by means of the survey is related to:

(i)   building characteristics (building location, surfaces, orientation, building envelope U-value, air exchange rate, occupancy profile, shadings);
(ii) building services system (heating system, cooling system, domestic hot water (DHW) plant, solar collectors, PV array);
(iii) appliances and devices (kitchen, refrigeration, washing, cleaning and ironing, lighting, audio/video, personal care, other equipment).

The questionnaire was built in the Excel environment based on Visual Basic for Applications. Real-time simulation of energy consumptions came from entered inputs. The in-house code adopted in previous studies [39–41] was validated by means of TRNSYS and EnergyPlus results comparison. The Heat Balance Method (HBM) was the base of the model implemented with a conduction finite difference (CondFD) solution algorithm [42]. Italian adoption of EU regulation was used for the calculation of heating, cooling, and DHW production efficiencies [43,44] for an expeditious assessment of the input data of questionnaire. Solar energy plants and their thermal and electrical outputs were also studied with technical norms in force [45]. Energy consumption of appliances and devices installed in the dwellings was based on their energy label [46–51]. Referring to lighting, further information was collected on the installed lamps, occupancy profile, and their location in the dwelling.

The simulation results were checked by comparison with energy bill values entered by the users. Further data sources were the Terna (Italian TSO) report [52] and ISTAT (National Statistics Institute) survey on residential energy consumption [53]. Given that, a first subdivision of loads in flexible and rigid ones was made. Five categories were identified:

(i)   Storable loads: charging and/or stopping load to create back-up for continuation;
(ii) Shiftable loads: impossible instant interruption but shift allowed over the time;
(iii) Curtailable loads: possible instant interruption and shift;
(iv) Non-curtailable loads: instant power not to be interrupted or shifted;
(v)   Self-generation: on-site power production acting on user-grid exchanges.

The main flexible loads in dwellings are "storable loads," i.e., heating, cooling, domestic hot water when equipped with battery or water tank and "shiftable loads," i.e., laundry, dishwasher, tumble dryer, vacuum cleaner, stove. The loads were identified by means of gathering surveys filled in for 419 dwellings from students of Faculty of Architecture at Sapienza University of Rome.

The yearly PEC primary energy consumptions equation in kWh/y reads as:

$$PEC = \sum_i \sum_j Q_{i,j} \cdot f_{ren,j} + \sum_i \sum_j Q_{i,j} \cdot f_{nren,j}, \tag{1}$$

where:

- $Q_{i,j}$ is the energy demand for i use such as electricity, heating, cooling, or domestic hot water associated to the used supply j in kWh/y;
- f_j is the primary energy conversion factor depending on *ren* renewable or *nren* fossil supply.

The yearly emission E in kg/y reads as:

$$E = \sum_i \sum_j Q_{i,j} \cdot f_{CO2,\,j},\tag{2}$$

where:

- $Q_{i,j}$ is the energy demand for i use such as electricity, heating, cooling, or domestic hot water associated with the used supply j in kWh/y;
- f_j is the emission factor associated to the used fuel j in kg/kWh.

The considered KPIs were calculated on the base of number of occupants, building surface, and in accordance with the Italian building energy certification system [54]. As already mentioned, some of the typical indicators were not present due to the high correlation. However, for this reason it was easy to calculate its value by rearranging some of the other ones [55].

2.1. Building Envelope Retrofitting Measures

Energy retrofitting of the building envelope has positive effects on the dwelling energy performance leading to a reduction of energy demand both in the winter heating season and in the summer cooling one. Conventional solutions are often the cost-effective ones even if crucial indicators such as levelized cost of energy [56] are not applied to them due to their large spread.

Considering the status quo, five alternative scenarios of building envelope energy retrofitting were simulated: (i) the insulation of vertical opaque walls, (ii) floors, (iii) ceilings, and (iv) the replacement of windows; first these interventions were considered individually and, then, (v) all together, as reported in Table 1. In all the cases it was assumed that, following the intervention, the values of the transmittance of the retrofitted building component were equal to those indicated later in Table 4 for the "after 2015" period. This parameter is essential for energy evaluation, while it does not provide information about other kinds of performances such as acoustics [57].

Table 1. Energy retrofitting measures for the building envelope.

#	Building Component	Retrofitting Measure
#1	Vertical opaque walls	
#2	Roof covering	Insulation up to the values of transmittance shown in Table 4 for
#3	Lower floor	the period "after 2015"
#4	Windows	
#5	Global upgrading of the envelope	#1 + #2 + #3 + #4

A limitation of generalized extension and size of the retrofitting measures is due to the fact that several building typologies were surveyed and in some cases, such as an apartment, not all the surfaces can be retrofitted without involving the next apartment or the roof is actually not present if the apartment is located at ground floor. It entails a new share of available interventions as below:

- insulation of vertical opaque walls, sample consisting of 403 dwellings, i.e., 96.2% of the total;
- insulation of the roof covering, 174 homes, i.e., 41.5%;
- floor insulation, 134 homes, i.e., 34.1%;
- replacement of windows, 391 homes, i.e., 93.3%;
- redevelopment of the entire building envelope, 418 homes, i.e., 99.8%.

2.2. Heating, Cooling, and DHW Systems Upgrading

The energy upgrading of technological systems produces positive effects on the energy performance of the dwelling, leading to an increase in the average seasonal yields of the systems. They are dynamically computed according to UNI/TS 11300/2 [43] for heating systems production and their regulation efficiencies. While, for a heat pump and its A+++ version, the Coefficient Of Performance (COP) is calculated in compliance with [46]. Considering the status quo, five alternative scenarios for improving the efficiency of the heating, DHW, and cooling systems were simulated, as reported in Table 2. Three measures (#6, #7, #8) are the replacement of existing equipment by a more efficient one whereas two measures (#9, #10) involve the heat pump technology installation for heating and for the DHW preparation, with electrification of these services if they were gas-fired.

Table 2. Interventions of technological systems upgrading.

#	System	Existing System	Upgrading Intervention
#6	Heating	Traditional boiler	- Substitution with condensing boiler - Installation of temperature control devices for single room
		Heat pump	- Substitution with heat pump A+++ class - Installation of temperature control devices for single room
#7	Heating—HP	Traditional boiler or heat pump	- Substitution with heat pump A+++ class - Installation of temperature control devices for single room
#8	DHW	Traditional boiler	- Substitution with condensing boiler
		Electric water heater	- Substitution with heat pump water heater
#9	DHW—HP	Traditional boiler or electric water heater	- Substitution with heat pump water heater
#10	Cooling	Air conditioner	- Substitution with air conditioner A+++ class

A first assessment was done accounting for the separated interventions, although in many of the examined dwellings, space heating and DHW rely on the same heat generator. Indeed, a limitation of the system upgrading is done by the presence of the already most efficient one or the absence of the one to be upgraded, such as the case of dwellings not equipped with cooling systems. In detail, among the total dwellings option #6 is applicable to 409 houses, i.e., 97.6%; option #7 to 414 houses, i.e., 98.8%; option #8 to 380 houses, i.e., 90.7%; option #9 to 411 homes, i.e., 98.1% and, finally, option #10 only to 204 homes, 48.7%. This latter one is actually the case of the dwellings equipped with cooling systems, therefore, suitable for upgrading.

2.3. Combined Building Envelope Retrofitting and System Upgrading

Finally, the effects of a combination of the building envelope retrofitting and of the system upgrading were simulated. Specifically, the energy retrofitting of the whole envelope was considered, together with the new technological systems, as reported in Table 3. Five combined scenarios were built. Considering the most frequent system layouts of the surveyed dwellings, scenarios #11 and #12 can be considered representative of a typical intervention carried out on a dwelling with centralized heating system. DHW production by means of a micro-heat pump is scenario #13 representing when there is no opportunity to upgrade the space heating system. On the contrary, scenarios #14 and #15 can be considered representative of comprehensive intervention carried out on a dwelling with an independent heating system.

Table 3. Combination of building envelope retrofitting and system upgrading.

#	Intervention	Description
#11	Overall envelope + heating system	#5 + #6
#12	Overall envelope + heating system HP	#5 + #7
#13	Overall envelope + DHW HP	#5 + #9
#14	Overall envelope + heating + DHW	#5 + #6 + #8
#15	Overall envelope + heating HP + DHW HP	#5 + #7 + #9

3. Results and Discussion

3.1. Description of Dwellings and Energy Consumption Analysis

This section shows the analysis regarding frequency of the building main features. Figure 1a depicts the subdivision of the buildings according to the construction year, following the official classification used in the Italian population censuses [58]. A further division compared to the decades is done to account for law modifications in the field of building energy performance (years 2005, 2008, 2010, 2015). The data show that a wide part of the sample buildings belongs to before 1976, exactly the year of the first Italian law about energy saving.

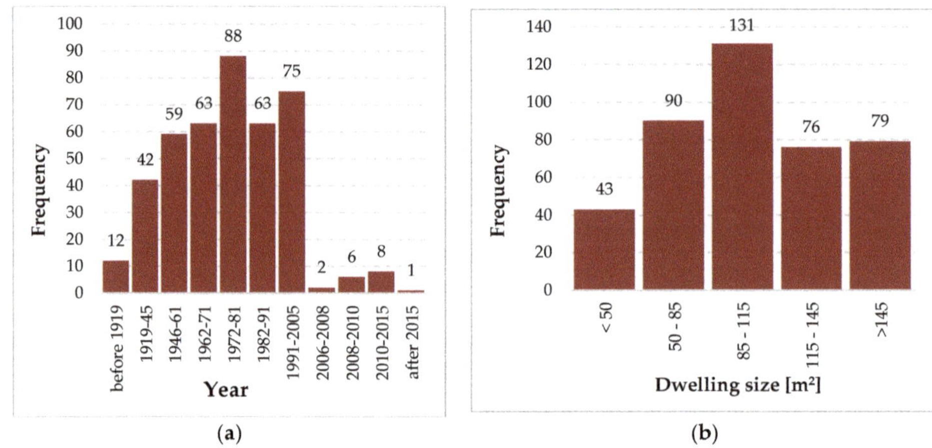

Figure 1. Dwellings' subdivision: (**a**) construction year; (**b**) size.

Moreover, Figure 1b shows the building frequency related to their size according to their five different dimensional classes specifically: small <50 m^2; small-medium 50–85 m^2; medium 85–115 m^2; medium-large 115–145 m^2; large >145 m^2. The average size of the apartments is equal to 112.4 m^2 and the most common class is the medium one. These different sizes are connected to the family number of components the dwelling was designed for. It has to be pointed out that a great part of the considered dwellings occupy only one floor, i.e., 360 equal to 86.5% while, among the remaining ones 37, i.e., 8.9%, occupy two floors and 19, i.e., 4.6%, three floors.

The data acquisition procedure does not include the introduction of the characteristics of the walls by the user, but there is the indication concerning the construction year and any refurbishment carried out. Table 4 shows the transmittance values of the buildings, depending on the construction year and according to the climate zone.

Reading the table, it is noteworthy that the current legislation provides a subdivision of the Italian territory into six climatic zones, according to the number of Degree Days (Zone A: DD ≤ 600; Zone B: 600 < DD ≤ 900; Zone C: 900 < DD ≤ 1400; Zone D: 1400 < DD ≤ 2100; Zone E: 2100 < DD ≤ 3000; Zone F: DD > 3000).

Table 4. Transmittance values depending on the construction year and the climate zone.

	Climate Zone	Construction Year										
		Before 1919	1919–1945	1946–1961	1962–1971	1972–1981	1982–1991	1991–2005	2006–2008	2008–2010	2010–2015	After 2015
Walls	A	1.30	1.20	1.20	1.20	1.10	1.00	0.90	0.86	0.69	0.62	0.62
	B	1.30	1.20	1.20	1.20	1.07	1.00	0.87	0.67	0.53	0.48	0.48
	C	1.30	1.20	1.20	1.20	1.03	0.98	0.83	0.56	0.44	0.40	0.40
	D	1.30	1.20	1.20	1.20	1.00	1.00	0.80	0.50	0.40	0.36	0.36
	E	1.23	1.13	1.13	1.13	0.94	0.94	0.76	0.47	0.38	0.34	0.34
	F	1.19	1.10	1.10	1.10	0.92	0.92	0.73	0.46	0.37	0.33	0.33
Roofs	A	1.30	1.30	1.30	1.30	1.20	1.07	0.95	0.55	0.42	0.38	0.33
	B	1.30	1.30	1.30	1.30	1.17	1.01	0.90	0.55	0.42	0.38	0.33
	C	1.30	1.30	1.30	1.30	1.12	0.96	0.85	0.55	0.42	0.38	0.33
	D	1.30	1.30	1.30	1.30	1.10	0.90	0.80	0.46	0.35	0.32	0.28
	E	1.22	1.22	1.22	1.22	1.03	0.84	0.75	0.43	0.33	0.30	0.26
	F	1.18	1.18	1.18	1.18	1.00	0.82	0.73	0.42	0.32	0.29	0.25
Floors	A	1.10	1.20	1.20	1.20	1.10	1.08	1.08	0.83	0.74	0.65	0.65
	B	1.10	1.20	1.20	1.20	1.03	0.92	0.92	0.63	0.56	0.49	0.49
	C	1.10	1.20	1.20	1.20	0.96	0.80	0.80	0.54	0.48	0.42	0.42
	D	1.10	1.20	1.20	1.20	0.90	0.60	0.60	0.46	0.41	0.36	0.36
	E	1.01	1.10	1.10	1.10	0.83	0.55	0.55	0.42	0.38	0.33	0.33
	F	0.98	1.07	1.07	1.07	0.80	0.53	0.53	0.41	0.36	0.32	0.32
Windows	A	5.20	5.10	5.00	5.00	5.00	5.00	5.00	5.00	5.00	4.60	4.03
	B	5.20	5.10	5.00	5.00	5.00	5.00	4.33	3.88	3.50	3.00	2.63
	C	5.20	5.10	5.00	5.00	4.86	4.86	3.81	3.36	3.03	2.60	2.28
	D	5.20	5.10	5.00	5.00	5.00	5.00	3.00	3.10	2.80	2.40	2.10
	E	4.77	4.68	4.58	4.58	4.58	4.58	2.75	2.84	2.57	2.20	1.93
	F	4.33	4.25	4.17	4.17	4.17	4.17	2.50	2.58	2.33	2.00	1.75

Table 5 points out the number of dwellings that have undergone renovations, divided according to year of construction and type of intervention. The most frequent refurbishment intervention is the replacement of windows, carried out in 197 dwellings (47.0% of the total) [59]. Air leakage entailing heat losses and acoustic discomfort are the main reasons of such intervention [60]. For the new installed building components, a transmittance value was set equal to the most common of the same period of installation.

Table 5. Retrofitted building components.

Construction Year of the Building	Retrofitted Building Component			
	Walls	Roofs	Floors	Windows
before 1919	0 (0%)	0 (0%)	0 (0%)	5 (41.7%)
1919–1945	2 (4.8%)	2 (4.8%)	3 (7.1%)	29 (69%)
1946–1961	6 (10.2%)	8 (13.6%)	1 (1.7%)	46 (78%)
1962–1971	8 (12.7%)	7 (11.1%)	3 (4.8%)	29 (46%)
1972–1981	13 (14.8%)	13 (14.8%)	5 (5.7%)	44 (50%)
1982–1991	6 (9.5%)	7 (11.1%)	4 (6.3%)	27 (42.9%)
1991–2005	16 (21.3%)	14 (18.7%)	9 (12%)	13 (17.3%)
2006–2008	0 (0%)	0 (0%)	0 (0%)	0 (0%)
2008–2010	2 (33.3%)	1 (16.7%)	0 (0%)	2 (33.3%)
2010–2015	4 (50%)	4 (50%)	3 (37.5%)	2 (25%)
after 2015	0 (0%)	0 (0%)	0 (0%)	0 (0%)
TOTAL	57 (13.6%)	56 (13.4%)	28 (6.7%)	197 (47%)

Regarding the HVAC systems, the analyzed dwellings are all equipped with a heating system and a DHW production system. Most of heating systems are autonomous (73.3%); gas is the most used energy vector in heating systems (98.8%) and in DHW preparation (85.4%). The majority of gas-fed systems provide both heating and DHW as shown in Figure 2.

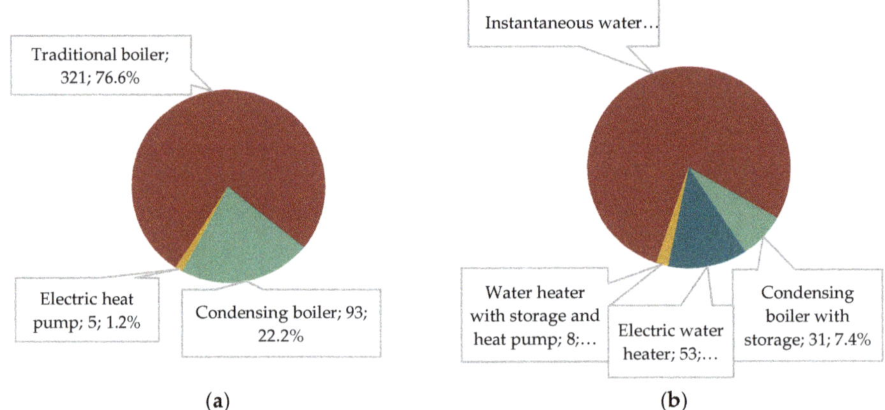

Figure 2. Types of heating systems: (**a**) space heating; (**b**) DHW.

Cooling systems are installed only in 207 dwellings (49.4%). They serve only a few rooms, as depicted by Figure 3a and they show high energy label due to their very recent installation, as Figure 3b reports.

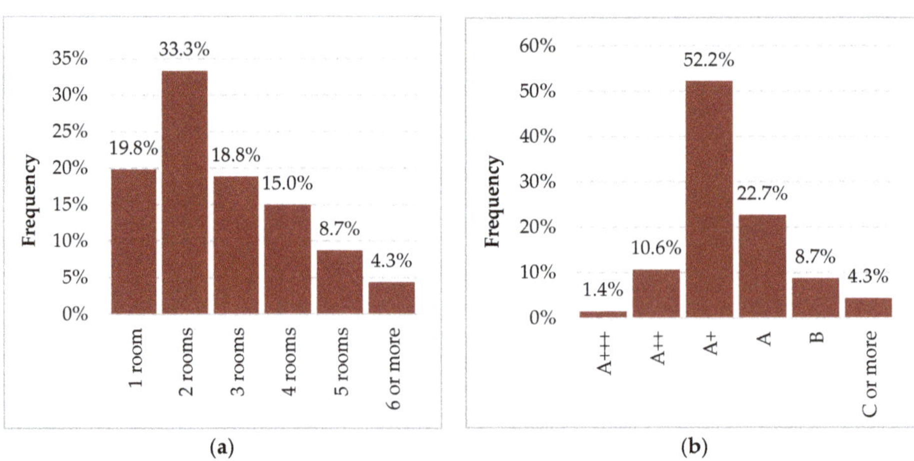

Figure 3. Cooling systems: (**a**) dwellings and number of air-conditioned rooms; (**b**) energy label.

In order to validate the simulations' outcomes from collected surveys, a comparison with energy bills was made. As a result of this comparison, a lower correlation between real and simulated gas consumption ($R^2 = 0.7764$) compared to the electricity one ($R^2 = 0.8977$) can be seen. The lower correlation relating to the use of gas depends on the greater uncertainty of occupants' profiles in dwellings and high incidence of this service on energy consumption.

The charts of Figures 4 and 5 show a calculation per unit of surface of the selected indicators: (i) primary energy consumption, (ii) local emissions, (iii) use of renewable energy, and (iv) flexible loads per class of sizes.

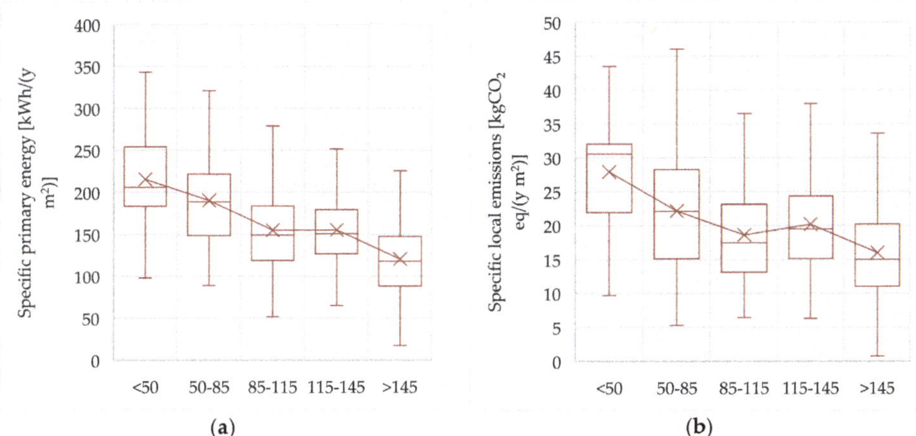

Figure 4. Key performance indicators: (**a**) specific primary energy; (**b**) specific local emissions.

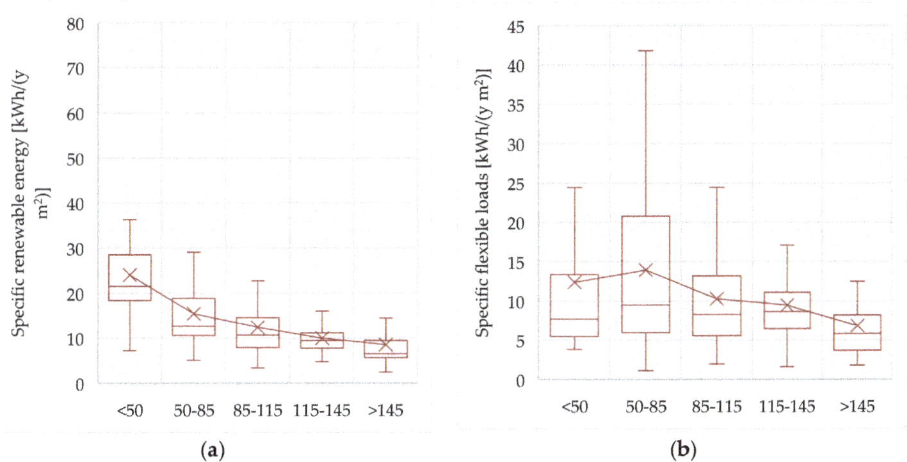

Figure 5. Key performance indicators: (**a**) specific renewable energy; (**b**) specific flexible loads.

At apartment size increases, the average value of primary energy consumption per unit of area decreases, from 215.0 kWh/m^{2}y (small dwelling) to 120.6 kWh/m^{2}y (large dwelling). Even the local emissions, in terms of equivalent CO_2, show a decreasing trend from 27.9 kg/m^{2}y (small dwelling) to 16.1 kg/m^{2}y (large dwelling). This is due to the lower ratio between surface and volume in the largest dwelling as well as lower specific occupancy rate. Regarding the medium-large dwellings, a slight variation due to the higher incidence of gas consumption for heating on the overall energy one is found. The larger houses with greater frequency are isolated. They have larger dispersing surfaces and, subsequently, heat losses [61].

Similarly, in Figure 5a, renewable energy consumption per unit of area decreases as the size of the apartment increases, from 24.1 kWh/m^{2}y (small dwelling) to 8.5 kWh/m^{2}y (large dwelling).

The consideration of another parameter is remarkable: the electrification degree. It is the ratio between the electricity consumption and the total primary energy one. It is on average 36.3% for the considered building stock. Next, the consumption of renewable energy is on average equal to 9.0%, depending largely on the renewable share of the power grid where they buy electricity from.

Even for flexible loads per unit of area, a decreasing trend is observed at increasing the dwelling surface, although not continuous, from 12.4 kWh/m^{2}y (small dwelling) to 6.9 kWh/m^{2}y (large dwelling). This is due the fact that different occupancy profiles occur since the small ones have higher rate of occupants' absence compared to the large one. In absolute terms, average flexible electric loads of the analyzed dwellings [54] are equal to 1043 kWh/y and they are largely shiftable loads (667 kWh/y) related to the use of washing machines, dishwashers, and dryers. Storable loads have a lower average magnitude (376 kWh/y), due to the low diffusion of electric heating and DHW systems and as a consequence of the low presence of cooling systems.

3.2. Building Envelope Retrofitting Measures

In all the retrofitting cases a reduction in heating and cooling consumption occurs. Its magnitude is linked to the relevance of the retrofitted building component on the energy consumption of the examined house, depending on the single case's geometric and thermo-physical parameters. Figure 6 shows the primary energy savings and the reduction of local emissions compared to the current situation. Evaluating the single interventions, the largest primary energy savings are linked to the interventions of isolation on the vertical opaque walls, on average equal to 16.3%, while, the smallest savings derive from the window ones, about 6.8%. This is due to the fact that, in many dwellings, the windows have been recently replaced. The intervention of insulation on ceilings and floors allows on average savings of 14.0% and 8.5%, respectively. However, the retrofitting of the whole building envelope can lead to a primary energy saving of about 30.8%. The trend of savings in local emissions is qualitatively similar to the primary energy one, although amplified: for vertical opaque walls a saving of 26.3% is achievable; for ceilings, 18.9%; for floors, 11.8%, and for fixtures, 11.5%. It leads to on average 47.6% in the case of the whole retrofitted building envelope. The reason for this amplification lies in the fact that, as seen, most of the heating systems are gas-fired and require local combustion to generate heat.

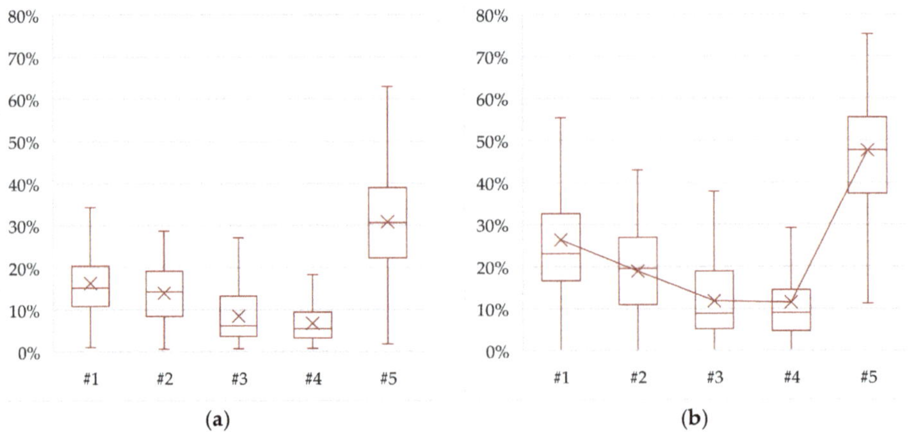

Figure 6. Resulted savings of (**a**) primary energy consumption; (**b**) local emissions.

Figure 7 shows the variations in the use of renewable energy and the number of flexible loads at each retrofitted building component.

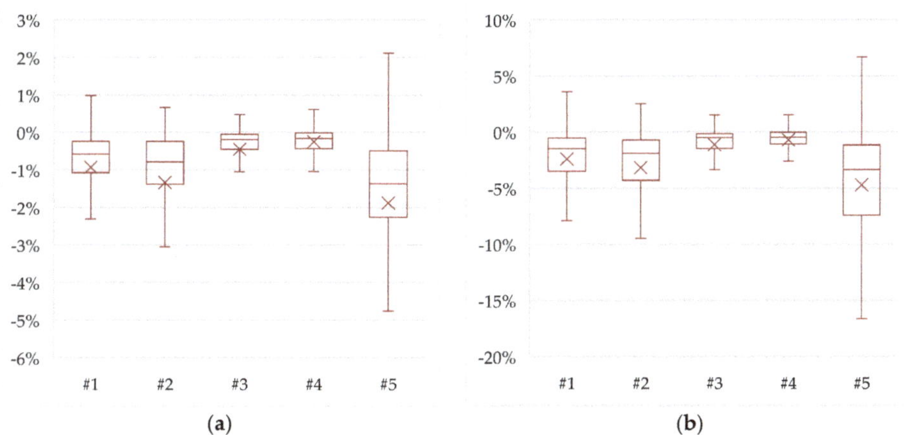

Figure 7. Resulted changes in (**a**) renewable energy use; (**b**) flexible loads.

The considered insulation interventions lead to a reduction in heating and cooling consumption, i.e., storable loads.

Moreover, the small variations are connected to the reduction of the electric consumption of the components of the heating system such as circulation pumps and fans, and to the reduction of cooling consumption, in the houses, if present.

Considering the whole building insulation intervention, the reduction of renewable energy use is less than 2% and the reduction of flexible loads is less than 5%.

Indeed, in the current plant configuration of Italian houses, a redevelopment intervention of the building envelope weakly affects the flexibility potential of the house. Yet, it entails valuable energy savings and reductions in local polluting emissions. In absolute terms, in the case of the whole building envelope retrofitting, flexible loads vary from 1043 to 1002 kWh/y.

3.3. Heating, Cooling, and DHW Systems Upgrading

In all the upgrading cases, a reduction in heating and cooling consumption occurs. Its magnitude is related to the relevance of the redeveloped element on the energy consumption of the analyzed dwelling, depending on the single case geometric and thermo-physical parameters.

Figure 8 shows the primary energy savings and the reduction of local emissions compared to the current situation. The largest savings in primary energy are observed for interventions on the heating system's heat generator, due to the high incidence of heating consumption on overall consumption [54]. For the upgrade option #6 the savings are on average equal to 6.5%, while for upgrade option #7 they are on average equal to 8.6%. Almost negligible savings, i.e., <1.5%, in all the other upgrading options occur.

Referring to the local emissions, the achievable reduction due to the installation of a heat pump as a heat generator and as the DHW production system are huge since they are 66.9% and 17.8%, respectively.

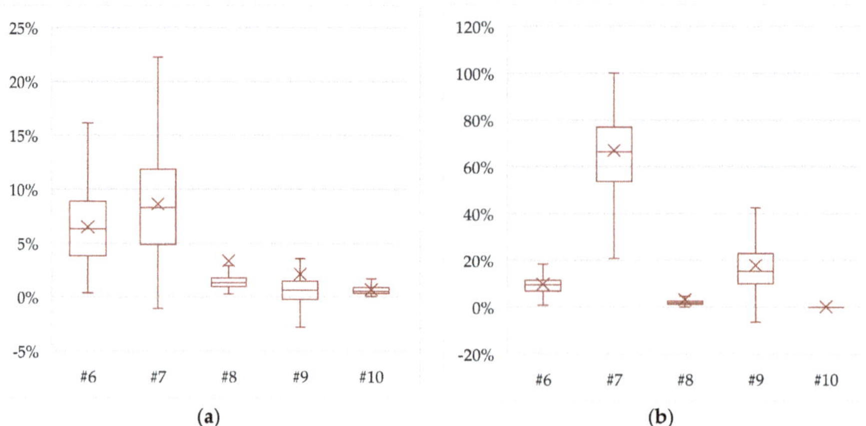

Figure 8. Resulted savings of (**a**) primary energy consumption; (**b**) local emissions.

Figure 9 shows the variations in the renewable energy use and the number of flexible loads at each system upgrade. Similar to the previous KPIs, the changes in those two thanks to the installation of heat pump are very significant (#7, #9). For the system upgrading #7, the increase in renewable energy use is on average 360% while, for the system upgrading #9, the variation is 45.4%. It is remarkable that for the other system upgrading the changes are very small: −1.2%, −2.1%, and −1.7% for #6, #8, and #10, respectively. The same behavior can be found for the changes in flexible loads. Indeed, system upgrading #7 and #9 imply their significant increases, about 147.0% and 80.1%, respectively. Then, small decreases occur for system upgrading #6, #8, and #10, i.e., −3.8%, −5.7%, and −4.8%. In absolute terms, the flexible loads in the system upgrading #7 reach 2111 kWh/y, whereas in #9, they reach 1443 kWh/y.

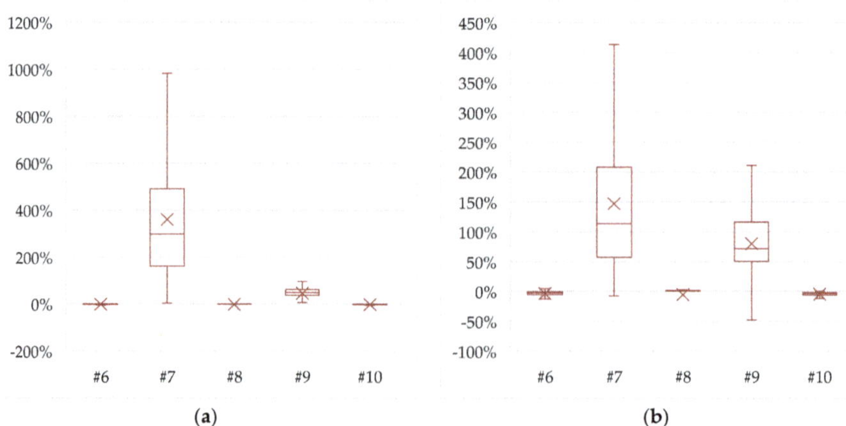

Figure 9. Resulted changes in (**a**) renewable energy use; (**b**) flexible loads.

The results of the simulations carried out confirm the usefulness of the heat pumps to increase the flexibility of the loads [62,63], as a basic element of a system that must necessarily include storage systems [64–66]. Anyway, the location of the storage system inside the dwellings remains to be explored. As a matter of fact, the DHW storage system is generally small [23] and easy to install inside the dwelling. Nevertheless, the storage system required for space heating is much larger, depending on the climate zone, the characteristics of the house, and the behavior of the occupants [67] together with needed preservation of architectural appearance when the building is considered historic or even listed [68].

3.4. Combined Building Envelope Retrofitting and System Upgrading

Finally, given the small changes observed in the previous section, the intervention to upgrade the cooling system was excluded.

Figure 10 shows the primary energy savings and the reduction of local emissions compared to the current situation. In terms of primary energy savings, the five combined scenarios are substantially equivalent, with savings ranging between 32.6%, in the case of #11, and 35.7%, in the case of #14.

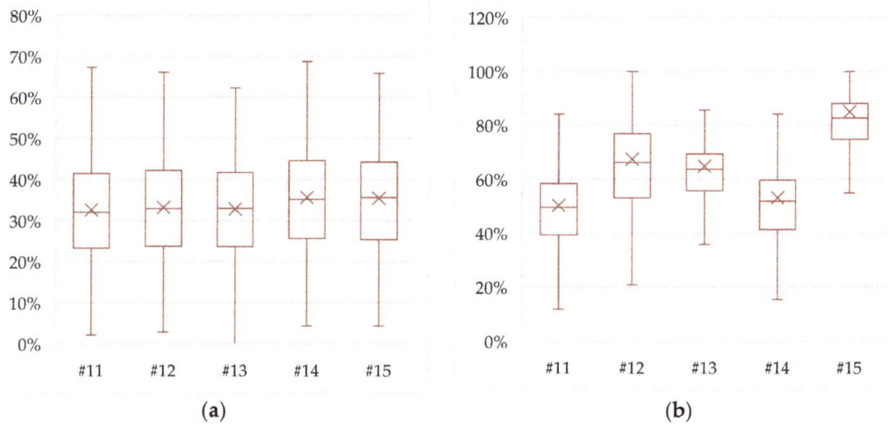

Figure 10. Resulted savings of (**a**) primary energy consumption; (**b**) local emissions.

Referring to local emissions, the achievable reductions are larger than 50% for all the combined scenarios, being able to reach 84.9% in the case of scenario #15, i.e., with complete electrification of space heating and DHW production.

Figure 11 shows the variations in the renewable energy use and the number of flexible loads at each combined retrofitting and system upgrading. As the outcome of system upgrading scenarios, the introduction of the heat pump as a heat generator significantly increase the renewable energy use, reaching on average values greater than 150% as in the case #15. Even in terms of flexible electric loads, a strong increase in the flexibility potential occurs, reaching 116.4% in the case of #15.

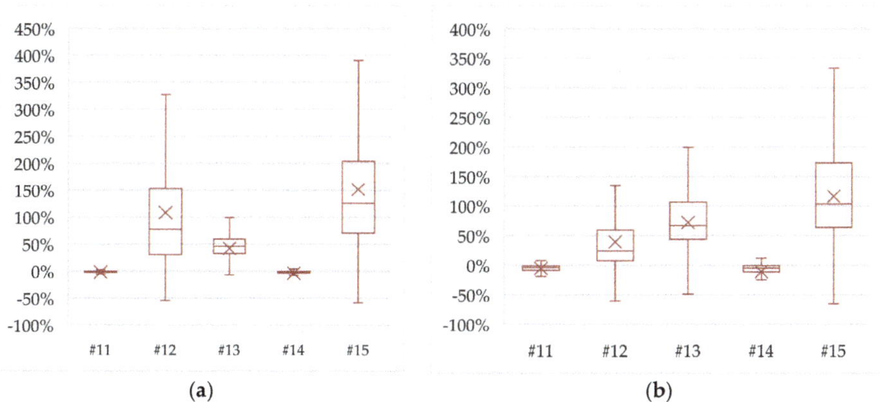

Figure 11. Resulted changes in (**a**) renewable energy use; (**b**) flexible loads.

In absolute terms, the value of flexible loads in scenarios #12, #13, and #15 reaches 1.329, 1.399, and 1.725 kWh/y, respectively.

These results are numerically lower than those found for scenario #7 where there was the installation of the single heat pump for space heating purposes. However, in this case, the lower increase in flexibility potential matches with positive values of all the other KPIs, i.e., primary energy savings, local emission reduction, and renewable energy use. Furthermore, the reduction of the heating demand has as a further positive aspect, an easier insertion of the storage system within the dwelling due to its new smaller required size [67].

3.5. Effects of the Proposed Measures

The changes deriving from the implementation of renovation measures are summarized in Table 6 reporting the values computed of the four KPIs.

Table 6. Summary of interventions and related changes.

#	Intervention	Resulted Savings		Resulted Changes	
		Primary Energy Consumption	Local Emissions	Renewable Energy Use	Flexible Loads
#1	Vertical opaque walls	16.3%	26.3%	−0.9%	−2.4%
#2	Roof covering	14.0%	18.9%	−1.3%	−3.2%
#3	Lower floor	8.5%	11.8%	−0.5%	−1.1%
#4	Windows	6.8%	11.5%	−0.3%	−0.7%
#5	Global upgrading of the envelope	30.8%	47.6%	−1.9%	−4.7%
#6	Heating	6.5%	9.7%	−1.2%	−3.8%
#7	Heating—HP	8.6%	66.9%	360.8%	147.0%
#8	DHW	3.4%	3.2%	−2.1%	−5.7%
#9	DHW—HP	2.1%	17.8%	45.4%	80.1%
#10	Cooling	0.7%	0.1%	−1.7%	−4.8%
#11	Overall envelope + heating system	32.6%	50.3%	−2.1%	−5.5%
#12	Overall envelope + heating system HP	33.2%	67.4%	107.9%	39.4%
#13	Overall envelope + DHW HP	32.8%	64.8%	41.9%	72.2%
#14	Overall envelope + heating + DHW	35.7%	53.1%	−3.9%	−10.8%
#15	Overall envelope + heating HP + DHW HP	35.5%	84.9%	151.6%	116.4%

Figure 12 depicts the electrification degree resulting from the 15 proposed interventions and combinations of them. In the status quo, the average electrification degree, i.e., the ratio between electrical loads and total ones, is 34%. Conversely, the highest value is reached in all the interventions apart from #10 which is 33.8% since the electric loads are reduced making the cooling supply more efficient. Interventions #5, #11, and #14 increase the mentioned parameter thanks the high reduction of heating consumption while interventions #7, #9, #12, #13, and #15 achieved the highest values switching the heating from fossil-based to electric-driven.

Figure 13a shows the flexible loads in percentage value in part while, in absolute values in Figure 13b. Strong changes are found only for the electrification of the heating systems by means of heat pumps for heating and/or for DHW. Those interventions are #7, #9, #12, #13, and #15.

Finally Figure 14a depicts how the frequency of flexible loads is before the intervention #0, and in Figure 14b after the comprehensive renovation #15. In the latter, 40% of dwellings show flexible loads lower than 1400 kWh/y. If the value 1800 kWh/y is considered, 60% of the analyzed residential building stock is under it. Only 32.2% of dwellings have more than 2000 kWh/y flexible loads.

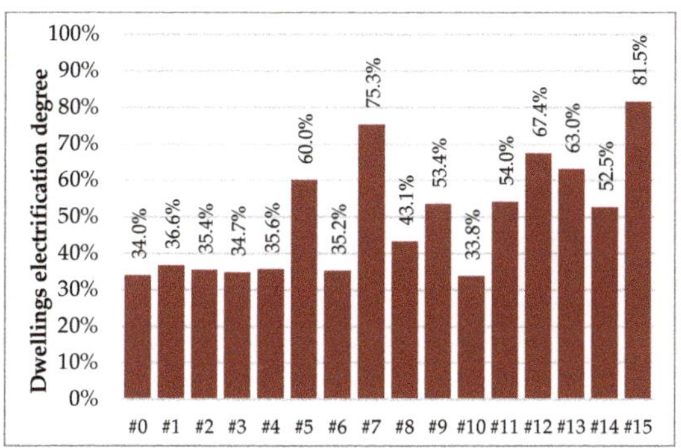

Figure 12. Dwellings' electrification degrees.

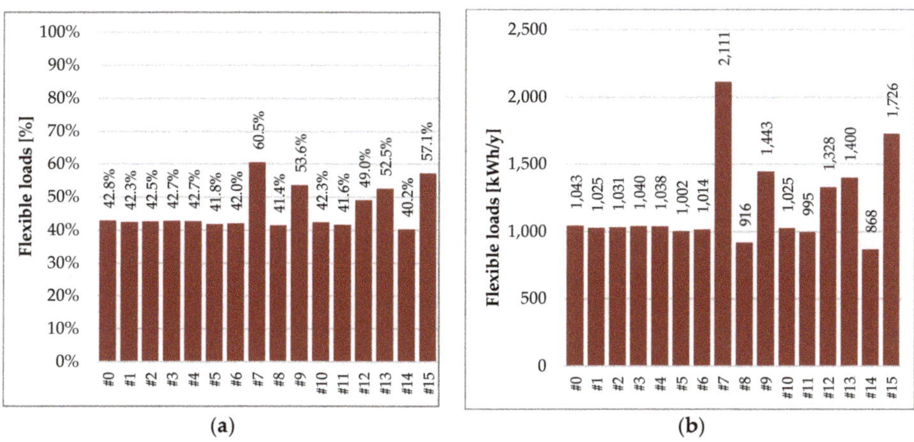

Figure 13. Flexible loads after renovation interventions. (**a**) Percentage values; (**b**) absolute values.

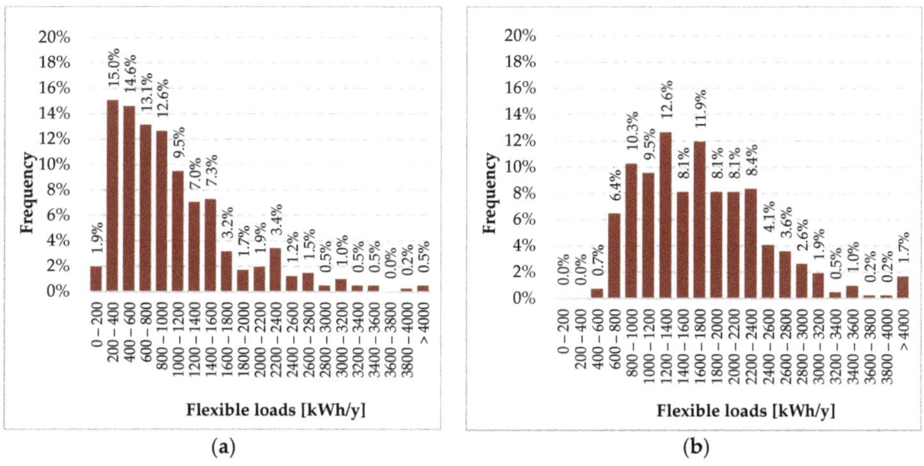

Figure 14. Flexible loads distribution. (**a**) Status quo #0; (**b**) after comprehensive intervention #15.

4. Conclusions

The European community has outlined its objectives in the field of environmental sustainability, indicating ambitious targets for reducing energy consumption, reducing emissions, and increasing the use of renewable energy sources. In this context, the study of energy retrofitting of existing buildings is of fundamental importance, as the building sector is responsible for more than 40% of total consumption. These interventions must be inserted in a new energy scenario, in which greater electrification of consumption and greater flexibility of demand will allow a wider integration of renewable energies. In this study, referring to the Italian situation, energy retrofitting interventions for the residential sector were analyzed, using a set of four Key Performance Indicators, i.e., primary energy consumption, renewable energy use, local polluting emissions, and flexible electrical loads. The interventions here analyzed are on the building envelope, on systems upgrading and, finally, a combination of them.

Considering the current situation, it was observed that:

- the Italian residential sector is not endowed with enough electrification; gas is the most widely used energy vector for space heating (98.8%) and DHW production (85.4%);
- only 49.4% of the dwellings surveyed are equipped with cooling systems;
- as the size of the apartment increases, primary energy consumption per unit area decreases; a similar trend is seen in the renewable energy use and local polluting emissions;
- for flexible loads per unit area, an overall decreasing trend occurs but with the exception of small-medium dwellings (50–85 m^2).
- As for the energy retrofitting interventions on the building envelope, it was observed that:
- all interventions involve a reduction in energy consumption and a reduction in polluting emissions;
- all the considered interventions involve a reduction in renewable energy use and a decrease in flexible loads.

With regard to the system upgrading, a clear difference was observed between gas-fed systems for space heating and DHW and electrified services by means of heat pump installation. In this case, a considerable increase in flexible loads and the renewable energy use occur with a strong reduction in local emissions due the increase of the power grid supply.

The results of the simulations carried out to evaluate the effects of a combined building envelope and system improvement are as summarized below:

- all combined interventions imply a reduction of primary energy consumption;
- a correlated reduction of local polluting emissions is found for gas-fed condensing boiler installation, while there are higher values of reduced local emissions in the case of electrification of space heating and DHW by means of heat pumps;
- all combined interventions entail a larger renewable energy use and a larger flexibility if heat pumps are to be installed;
- a total of 50% of the renovated dwellings in the #15 scenario show a flexibility between 1200 and 2400 kWh/y entailing the possibility to use it as a reference value for preliminary calculation of aggregated users.

It is noteworthy how the results of this study can offer a further KPI to evaluate the best retrofitting intervention, from an economic perspective, accounting for new electricity business models in the era of prosumers where flexibility is becoming the new performance of the built environment.

Author Contributions: The authors equally contributed to this paper.

Funding: This research received no external funding.

Conflicts of Interest: The authors declare no conflict of interest.

References

1. European Commission. *A Clean Planet for All—A European Strategic Long-Term Vision for a Prosperous, Modern, Competitive and Climate Neutral Economy*; European Commission: Brussels, Belgium, 2018.

2. Eurostat Database—Eurostat. Available online: https://ec.europa.eu/eurostat/data/database (accessed on 4 April 2019).

3. Piras, G.; Pini, F.; Garcia, D.A. Correlations of PM10 concentrations in urban areas with vehicle fleet development, rain precipitation and diesel fuel sales. *Atmospheric Pollut. Res.* **2019**, *10*, 1165–1179. [CrossRef]

4. Aste, N.; Manfren, M.; Marenzi, G. Building Automation and Control Systems and performance optimization: A framework for analysis. *Renew. Sustain. Energy Rev.* **2017**, *75*, 313–330. [CrossRef]

5. Nezhad, M.M.; Groppi, D.; Rosa, F.; Piras, G.; Cumo, F.; Garcia, D.A. Nearshore wave energy converters comparison and Mediterranean small island grid integration. *Sustain. Energy Technol. Assess.* **2018**, *30*, 68–76.

6. Tronchin, L.; Manfren, M.; Nastasi, B. Energy analytics for supporting built environment decarbonisation. *Energy Procedia* **2019**, *157*, 1486–1493. [CrossRef]

7. Nastasi, B. *"Hydrogen Policy, Market and R&D Projects" in Solar Hydrogen Production*; Calise, F., D'Accadia, M.D., Santarelli, M., Lanzini, A., Ferrero, D., Eds.; Elsevier: Amsterdam, The Netherlands, 2019; ISBN 9780128148532.

8. Zappa, W.; Junginger, M.; van den Broek, M. Is a 100% renewable European power system feasible by 2050? *Appl. Energy* **2019**, *233–234*, 1027–1050. [CrossRef]

9. De Santoli, L.; Mancini, F.; Garcia, D.A. A GIS-based model to assess electric energy consumptions and usable renewable energy potential in Lazio region at municipality scale. *Sustain. Cities Soc.* **2019**, *46*, 101413. [CrossRef]

10. Adhikari, R.; Aste, N.; Manfren, M.; Adhikari, R. Multi-commodity network flow models for dynamic energy management—Smart Grid applications. *Energy Procedia* **2012**, *14*, 1374–1379. [CrossRef]

11. Hossain, M.; Madlool, N.; Rahim, N.; Selvaraj, J.; Pandey, A.; Khan, A.F. Role of smart grid in renewable energy: An overview. *Renew. Sustain. Energy Rev.* **2016**, *60*, 1168–1184. [CrossRef]

12. Rahmani-Andebili, M. Modeling nonlinear incentive-based and price-based demand response programs and implementing on real power markets. *Electr. Power Syst. Res.* **2016**, *132*, 115–124. [CrossRef]

13. Aghajani, G.; Shayanfar, H.; Shayeghi, H. Demand side management in a smart micro-grid in the presence of renewable generation and demand response. *Energy* **2017**, *126*, 622–637. [CrossRef]

14. Tronchin, L.; Manfren, M.; James, P.A. Linking design and operation performance analysis through model calibration: Parametric assessment on a Passive House building. *Energy* **2018**, *165*, 26–40. [CrossRef]

15. De Santoli, L.; Garcia, D.A.; Groppi, D.; Bellia, L.; Palella, B.I.; Riccio, G.; Cuccurullo, G.; Ambrosio, F.R.D.; Stabile, L.; Dell'Isola, M.; et al. A General Approach for Retrofit of Existing Buildings Towards NZEB: The Windows Retrofit Effects on Indoor Air Quality and the Use of Low Temperature District Heating. In Proceedings of the 2018 IEEE International Conference on Environment and Electrical Engineering and 2018 IEEE Industrial and Commercial Power Systems Europe (EEEIC/I&CPS Europe), Palermo, Italy, 12–15 June 2018; pp. 1–6.

16. Castellani, B.; Morini, E.; Nastasi, B.; Nicolini, A.; Rossi, F. Small-Scale Compressed Air Energy Storage Application for Renewable Energy Integration in a Listed Building. *Energies* **2018**, *11*, 1921. [CrossRef]

17. Bartusch, C.; Wallin, F.; Odlare, M.; Vassileva, I.; Wester, L. Introducing a demand-based electricity distribution tariff in the residential sector: Demand response and customer perception. *Energy Policy* **2011**, *39*, 5008–5025. [CrossRef]

18. Wang, J.; Li, P.; Fang, K.; Zhou, Y. Robust Optimization for Household Load Scheduling with Uncertain Parameters. *Appl. Sci.* **2018**, *8*, 575. [CrossRef]

19. Koch, A.; Girard, S.; McKoen, K. Towards a neighbourhood scale for low- or zero-carbon building projects. *Build. Res. Inf.* **2012**, *40*, 527–537. [CrossRef]

20. Corbin, C.D.; Henze, G.P. Predictive control of residential HVAC and its impact on the grid. Part II: simulation studies of residential HVAC as a supply following resource. *J. Build. Perform. Simul.* **2017**, *10*, 365–377. [CrossRef]

21. Li, W.; Yang, L.; Ji, Y.; Xu, P. Estimating demand response potential under coupled thermal inertia of building and air-conditioning system. *Energy Build.* **2019**, *182*, 19–29. [CrossRef]

22. Foteinaki, K.; Li, R.; Heller, A.; Rode, C. Heating system energy flexibility of low-energy residential buildings. *Energy Build.* **2018**, *180*, 95–108. [CrossRef]

23. Balint, A.; Kazmi, H. Determinants of energy flexibility in residential hot water systems. *Energy Build.* **2019**, 286–296. [CrossRef]

24. Issi, F.; Kaplan, O. The Determination of Load Profiles and Power Consumptions of Home Appliances. *Energies* **2018**, *11*, 607. [CrossRef]

25. Ruzzenenti, F.; Bertoldi, P. *Energy Conservation Policies in the Light of the Energetics of Evolution*; Springer Science and Business Media LLC: Berlin/Heidelberg, Germany, 2017; Volume 493, pp. 147–167.

26. Tzeiranaki, S.T.; Bertoldi, P.; Diluiso, F.; Castellazzi, L.; Economidou, M.; Labanca, N.; Serrenho, T.R.; Zangheri, P. Analysis of the EU Residential Energy Consumption: Trends and Determinants. *Energies* **2019**, *12*, 1065. [CrossRef]

27. Wang, B.; Xia, X.; Zhang, J. A multi-objective optimization model for the life-cycle cost analysis and retrofitting planning of buildings. *Energy Build.* **2014**, *77*, 227–235. [CrossRef]

28. Fan, Y.; Xia, X. A multi-objective optimization model for energy-efficiency building envelope retrofitting plan with rooftop PV system installation and maintenance. *Appl. Energy* **2017**, *189*, 327–335. [CrossRef]

29. Wu, R.; Mavromatidis, G.; Orehounig, K.; Carmeliet, J. Multiobjective optimisation of energy systems and building envelope retrofit in a residential community. *Appl. Energy* **2017**, *190*, 634–649. [CrossRef]

30. Penna, P.; Prada, A.; Cappelletti, F.; Gasparella, A. Multi-objectives optimization of Energy Efficiency Measures in existing buildings. *Energy Build.* **2015**, *95*, 57–69. [CrossRef]

31. Asadi, E.; Da Silva, M.G.; Antunes, C.H.; Dias, L.; Da Silva, M.C.G. Multi-objective optimization for building retrofit strategies: A model and an application. *Energy Build.* **2012**, *44*, 81–87. [CrossRef]

32. Guardigli, L.; Bragadin, M.A.; Della Fornace, F.; Mazzoli, C.; Prati, D. Energy retrofit alternatives and cost-optimal analysis for large public housing stocks. *Energy Build.* **2018**, *166*, 48–59. [CrossRef]

33. Dall'O', G.; Galante, A.; Pasetti, G. A methodology for evaluating the potential energy savings of retrofitting residential building stocks. *Sustain. Cities Soc.* **2012**, *4*, 12–21. [CrossRef]

34. De Santoli, L.; Mancini, F.; Nastasi, B.; Ridolfi, S. Energy retrofitting of dwellings from the 40's in Borgata Trullo-Rome. *Energy Procedia* **2017**, *133*, 281–289. [CrossRef]

35. Mancini, F.; Salvo, S.; Piacentini, V. Issues of Energy Retrofitting of a Modern Public Housing Estates: The 'Giorgio Morandi' Complex at Tor Sapienza, Rome, 1975–1979. *Energy Procedia* **2016**, *101*, 1111–1118. [CrossRef]

36. Niemelä, T.; Kosonen, R.; Jokisalo, J. Energy performance and environmental impact analysis of cost-optimal renovation solutions of large panel apartment buildings in Finland. *Sustain. Cities Soc.* **2017**, *32*, 9–30. [CrossRef]

37. Ortiz, J.I.; Casas, A.F.; Salom, J.; Soriano, N.G.; Casas, P.F.I.; Ferrà, J.A.O. Cost-effective analysis for selecting energy efficiency measures for refurbishment of residential buildings in Catalonia. *Energy Build.* **2016**, *128*, 442–457. [CrossRef]

38. Valančius, K.; Vilutienė, T.; Rogoža, A. Analysis of the payback of primary energy and CO2 emissions in relation to the increase of thermal resistance of a building. *Energy Build.* **2018**, *179*, 39–48. [CrossRef]

39. Mancini, F.; Cecconi, M.; De Sanctis, F.; Beltotto, A. Energy Retrofit of a Historic Building Using Simplified Dynamic Energy Modeling. *Energy Procedia* **2016**, *101*, 1119–1126. [CrossRef]

40. De Santoli, L.; Mancini, F.; Clemente, C.; Lucci, S. Energy and technological refurbishment of the School of Architecture Valle Giulia, Rome. *Energy Procedia* **2017**, *133*, 382–391. [CrossRef]

41. Mancini, F.; Clemente, C.; Carbonara, E.; Fraioli, S. Energy and environmental retrofitting of the university building of Orthopaedic and Traumatological Clinic within Sapienza Città Universitaria. *Energy Procedia* **2017**, *126*, 195–202. [CrossRef]

42. ASHRAE. *Handbook—Fundamentals (SI Edition)*; ASHRAE: Atlanta, GA, USA, 2017; ISBN 6785392187.

43. UNI Ente Italiano di Normazione Italian technical standard UNI/TS 11300-2:2019—Evaluation of Primary Energy Need and of System Efficiencies for Space Heating, Domestic Hot Water Production, Ventilation and Lighting for Non-Residential Buildings. Available online: http://store.uni.com/catalogo/index.php/uni-ts-11300-2-2019.html?josso_back_to=http://store.uni.com/josso-security-check.php&josso_cmd=login_optional&josso_partnerapp_host=store.uni.com (accessed on 18 May 2019).

44. UNI Ente Italiano di Normazione Italian technical standard UNI/TS 11300-3:2010—Evaluation of Primary Energy and System Efficiencies for Space Cooling. Available online: http://store.uni.com/catalogo/index.php/uni-ts-11300-3-2010.html (accessed on 18 May 2019).

45. UNI Ente Italiano di Normazione Italian technical standard UNI/TS 11300-4:2012—Renewable Energy and Other Generation Systems for Space Heating and Domestic Hot Water Production. Available online: http://store.uni.com/catalogo/index.php/uni-ts-11300-4-2012.html (accessed on 18 May 2019).

46. European Commission. *COMMISSION DELEGATED REGULATION (EU) No 626/2011*; European Commission: Brussels, Belgium, 2011.
47. European Commission. *COMMISSION DELEGATED REGULATION (EU) No 1060/2010*; European Commission: Brussels, Belgium, 2010.
48. European Commission. *COMMISSION DELEGATED REGULATION (EU) No 1061/2010*; European Commission: Brussels, Belgium, 2010.
49. European Commission. *COMMISSION DELEGATED REGULATION (EU) No 392/2012*; European Commission: Brussels, Belgium, 1998.
50. European Commission. *COMMISSION DELEGATED REGULATION (EU) No 1059/2010*; European Commission: Brussels, Belgium, 2010.
51. European Commission. *COMMISSION DELEGATED REGULATION (EU) No 1062/2010*; European Commission: Brussels, Belgium, 2010.
52. Terna. Consumi Energia Elettrica per Settore Merceologico. Available online: http://www.terna.it/default/Home/SISTEMA_ELETTRICO/statistiche/consumi_settore_merceologico.aspx (accessed on 4 April 2019).
53. ISTAT. *I Consumi Energetici Delle Famiglie*; ISTAT: Rome, Italy, 2014.
54. Mancini, F.; Basso, G.L.; De Santoli, L. Energy Use in Residential Buildings: Characterisation for Identifying Flexible Loads by Means of a Questionnaire Survey. *Energies* **2019**, *12*, 2055. [CrossRef]
55. Kylili, A.; Fokaides, P.A.; Jimenez, P.A.L. Key Performance Indicators (KPIs) approach in buildings renovation for the sustainability of the built environment: A review. *Renew. Sustain. Energy Rev.* **2016**, *56*, 906–915. [CrossRef]
56. Zupone, G.L.; Amelio, M.; Barbarelli, S.; Florio, G.; Scornaienchi, N.M.; Cutrupi, A. Levelized Cost of Energy: A First Evaluation for a Self Balancing Kinetic Turbine. *Energy Procedia* **2015**, *75*, 283–293. [CrossRef]
57. Tronchin, L. On the acoustic efficiency of road barriers: The Reflection Index. *Int. J. Mech.* **2013**, *7*, 318–326.
58. ISTAT. Edifici Residenziali. Available online: http://dati-censimentopopolazione.istat.it/Index.aspx?DataSetCode=DICA_EDIFICIRES (accessed on 4 April 2019).
59. Piras, G.; Pennacchia, E.; Barbanera, F.; Cinquepalmi, F. The use of local materials for low-energy service buildings in touristic island: The case study of Favignana island. In Proceedings of the 2017 IEEE International Conference on Environment and Electrical Engineering and 2017 IEEE Industrial and Commercial Power Systems Europe (EEEIC/I&CPS Europe), Milan, Italy, 6–9 June 2017; pp. 1–4.
60. Tronchin, L.; Coli, V.L. Further Investigations in the Emulation of Nonlinear Systems with Volterra Series. *J. Audio Eng. Soc.* **2015**, *63*, 671–683. [CrossRef]
61. Garcia, D.A.; Cumo, F.; Pennacchia, E.; Pennucci, V.S.; Piras, G.; De Notti, V.; Roversi, R.; Brebbia, C.A. Assessment of a urban sustainability and life quality index for elderly. *Int. J. Sustain. Dev. Plan.* **2017**, *12*, 908–921. [CrossRef]
62. Hewitt, N.J. Heat pumps and energy storage—The challenges of implementation. *Appl. Energy* **2012**, *89*, 37–44. [CrossRef]
63. Chen, Y.; Xu, P.; Gu, J.; Schmidt, F.; Li, W. Measures to improve energy demand flexibility in buildings for demand response (DR): A review. *Energy Build.* **2018**, *177*, 125–139. [CrossRef]
64. Moreno, P.; Castell, A.; Solé, C.; Zsembinszki, G.; Cabeza, L.F. PCM thermal energy storage tanks in heat pump system for space cooling. *Energy Build.* **2014**, *82*, 399–405. [CrossRef]
65. Floss, A.; Hofmann, S. Optimized integration of storage tanks in heat pump systems and adapted control strategies. *Energy Build.* **2015**, *100*, 10–15. [CrossRef]
66. Hesaraki, A.; Holmberg, S.; Haghighat, F. Seasonal thermal energy storage with heat pumps and low temperatures in building projects—A comparative review. *Renew. Sustain. Energy Rev.* **2015**, *43*, 1199–1213. [CrossRef]
67. Alimohammadisagvand, B.; Jokisalo, J.; Kilpeläinen, S.; Ali, M.; Sirén, K. Cost-optimal thermal energy storage system for a residential building with heat pump heating and demand response control. *Appl. Energy* **2016**, *174*, 275–287. [CrossRef]
68. Garcia, D.A.; Di Matteo, U.; Cumo, F. Selecting Eco-Friendly Thermal Systems for the "Vittoriale Degli Italiani" Historic Museum Building. *Sustainbility* **2015**, *7*, 12615–12633. [CrossRef]

Article

Development of Demand Response Energy Management Optimization at Building and District Levels Using Genetic Algorithm and Artificial Neural Network Modelling Power Predictions

Nikos Kampelis [1],*, Elisavet Tsekeri [1], Dionysia Kolokotsa [1], Kostas Kalaitzakis [2], Daniela Isidori [3] and Cristina Cristalli [3]

[1] Energy Management in the Built Environment Research Lab, Environmental Engineering School, Technical University of Crete, Technical University Campus, Kounoupidiana, GR 73100 Chania, Greece; etsekeri@isc.tuc.gr (E.T.); dkolokotsa@enveng.tuc.gr (D.K.)

[2] Electric Circuits and Renewable Energy Sources Laboratory, Technical University of Crete, GR 73100 Chania, Greece; koskal@elci.tuc.gr

[3] Research for Innovation, AEA srl. via Fiume 16, IT 60030 Angeli di Rosora, Marche, Italy; d.isidori@loccioni.com (D.I.); c.cristalli@loccioni.com (C.C.)

* Correspondence: nkampelis@isc.tuc.gr; Tel.: +30-28210-37765; Fax: +30-28210-37846

Received: 25 September 2018; Accepted: 26 October 2018; Published: 1 November 2018

Abstract: Demand Response (DR) is a fundamental aspect of the smart grid concept, as it refers to the necessary open and transparent market framework linking energy costs to the actual grid operations. DR allows consumers to directly or indirectly participate in the markets where energy is being exchanged. One of the main challenges for engaging in DR is associated with the initial assessment of the potential rewards and risks under a given pricing scheme. In this paper, a Genetic Algorithm (GA) optimisation model, using Artificial Neural Network (ANN) power predictions for day-ahead energy management at the building and district levels, is proposed. Individual building and building group analysis is conducted to evaluate ANN predictions and GA-generated solutions. ANN-based short term electric power forecasting is exploited in predicting day-ahead demand, and form a baseline scenario. GA optimisation is conducted to provide balanced load shifting and cost-of-energy solutions based on two alternate pricing schemes. Results demonstrate the effectiveness of this approach for assessing DR load shifting options based on a Time of Use pricing scheme. Through the analysis of the results, the practical benefits and limitations of the proposed approach are addressed.

Keywords: demand response; artificial neural network; power predictions; energy management; genetic algorithm; optimisation; microgrid; smart grid

---

1. Introduction

Preparation for the transition from conventional power grids to next generation, so-called "smart" grids, is a worldwide trend nowadays. The goal for stakeholders in the domains of operations, generation, transmission, distribution, and service provision [1] is to offer more and higher quality services while improving operational capabilities, flexibility, and energy efficiency. In this context, a higher-level utilisation of smart grid resources is targeted by grid modernisation and enhanced dispersed dynamic measurements at local, regional, and wider levels. Various forms of communication equipment and protocols allow smart metering, monitoring, and controls in an interoperable unified system often described as Advanced Metering Infrastructure (AMI).

Smart metering and AMI are widely recognized as a necessity for the reliable and fast exchange of data in smart grids [2]. It is expected that nodal analysis of power measurements in the power grid will provide valuable information for utilities to control multi-directional flows of energy and improve dispatching, addressing vulnerabilities and constraints. In this sense, it is foreseen that a variety of technological solutions will emerge to balance the high volatility and power quality issues of the miscellaneous intermittent loads and renewable energy sources.

On the market side, reforms are required to leverage innovation in services and new business models which will upgrade existing operations. In this context, Demand Response constitutes a variety of services which have transformed the electric grid and energy markets operations during the past decades. Significant progress has been made in the US, where DR programs have been designed and implemented for years, and span across the full range of dispatchable (reliability, economic) and non-dispatchable (time sensitive pricing; ToU, CPP, RTP) demand side management options [3]. Demand side management is a valuable prospect for consumers and utilities—if used properly—for the use of assets and to decrease losses in transmission and distribution, as well as reducing avoidable costs. In this context, DR, along with the demand-side management of distributed energy resources, expand the boundaries for near future scientific and technological advances.

In the European Union, the Energy Efficiency Directive (EED), 2012/27/EU foresees the elimination of barriers for Demand Response (DR) in balancing and ancillary services markets [4]. Among the EU Member States (MS), considering the progress in DR, Belgium, France, Ireland, and UK, are in the leading group. Significant steps have also been taken in this direction by Germany, the Nordic countries, the Netherlands, and Austria. Generally, DR programs are differentiated (a) explicitly, i.e., where DR participants transact directly in the energy market, and (b) implicitly, i.e., where participation through a third party is facilitated [5].

Furthermore, Open Automated Demand Response (OpenADR) is a well-established protocol defining various deployment scenarios for facilitating DR programs and measures [6]. The overall framework of smart grids with regards to DR is presented and analyzed by Siano in [7]. Important aspects are defined, and a description of the possibilities created by DR for utilities and customers are analyzed. Load curtailment, shifting energy consumption, and using onsite energy generation, thus reducing the dependence on the main grid, are the main mechanisms for customers to participate in DR. Customer participation in wholesale markets via intermediaries, such as curtailment service providers (CSP), aggregators, or retail customers (ARC), demand response providers (DRPs), or local distribution companies, is documented in [7]. Moreover, a review of DR and smart grids with respect to the potential benefits and enabling technologies is provided. Considering system operation, contingency issues can be dealt with through DR implementation, resulting in a reduction of electrical consumption at critical hours, and avoiding serious impacts due to failure of power services provision. Considering energy efficiency, it is ascertained that effective management of aggregated loads can lead to a reduction of the overall cost of energy, due to the reduction and operating-time-shortening of conventional power generation equipment. Avoiding network upgrades at the local level, or postponing investments in new capacity, reserves or peaking units at system level, is another important potential benefit linked to high level implementation of DR. Modelling of incentive-based DR focusing on interruptible/curtailable service and capacity market programs is investigated by Aalami et al. in [8]. Price elasticity of demand, and a customer benefit function, are used to develop an economic model. Several scenarios are simulated and evaluated based on their value according to different strategies and performance with respect to improvement of the load curve (peak reduction, load factor, peak to valley), benefit of customers, and reduction of energy consumption.

Wholesale electricity market design considerations with regards to major challenges, aiming at increasing renewable energy penetration, are explored in [9]. Various dynamic energy pricing models have been proposed to compensate for market uncertainty and risks [10,11]. A residential DR based on adaptive consumption pricing is proposed by Haider [12], allowing utilities to manage aggregate load, and customers to lower their energy consumption. The proposed pricing scheme adapts energy costs to customers' consumption levels, thus encouraging active enrolment in the DR program. Cost and comfort optimisation of load scheduling under different pricing schemes has been investigated using various techniques including linear, convex, PSO, MINLP [13]. Furthermore, technology readiness, opportunities, and requirements for the deployment of DR in buildings and blocks of buildings is addressed by Crosbie et al. in [14,15].

On the other hand, buildings worldwide are responsible for over 40% of total energy consumption, gas emissions, and global warming [16]. The role of smart grids for near- and zero-energy building communities is investigated by researchers to test new approaches, identify critical aspects, and tackle challenges emerging when dealing with design and operational problems [17,18]. On the demand side, a wide variety of developed scientific tools influence the dynamics of advances in energy performance and energy management in buildings [19–22]. Such tools are embedded in data monitoring applications, such as innovative web-based energy management platforms [23,24] to enable improved analysis, decision making, and dynamic controls. Moving from Building Energy Management Systems (BEMS) [25,26] to District Energy Management Systems (DEMS) [27] entails the dynamic exchange and hierarchical processing of data streams between various components and systems, as in the Internet of Things (IoT) paradigm [28,29]. Various techniques and tools have been investigated for dealing with challenges in various fields pertaining to smart grids: smart metering data analysis and dynamic processing [30], power demand forecasting [31,32], Distributed Energy Resources (DER) management optimisation [33], users' engagement [34], etc.

In addition, Hybrid Renewable Energy Systems (HRES) have been implemented in various configurations to combine two or more renewable and non-renewable sources in order to deal with the intermittency of renewable energy sources, such as solar or wind. HRES have important attributes which make them increasingly attractive as alternatives to conventional fossil fuel energy sources in numerous applications [35–38]. Aligned with HRES, the concept of the microgrid as a semi-autonomous system of increased flexibility and manageable energy resources, such as renewable energy generation, storage, backup systems and flexible demand, is of particular importance when it comes to supporting grid stability and decentralized control [39]. A comprehensive critical review on the energy management systems of microgrids is conducted by Zia et al. in [40], with reference to the level of maturity of real world applications. Communication issues, control technologies and architectures, deployment costs, energy management strategies, optimisation, objectives and limitations, are addressed. An auto-configuration function using a multi-agent approach is proposed in [41] to establish automatic connection or disconnection of DER at microgrid level, capable of dealing with system faults and re-optimising the new configuration as necessary. Unsymmetrical and ground faults analysis in microgrids distribution systems is proposed by Ou in [42,43]. Hirsch et al. in [44] surveyed technologies and key drivers of microgrid implementation and research, at international level. Reported drivers in this context include extreme weather related concerns, cascading outages, cyber and physical attacks, deferral of infrastructure expansion costs, reduced line losses, efficiency improvements, savings, responsiveness, balancing loads, RE generation, etc. In [45], the authors present a residential microgrid day-ahead planning approach to accommodate appliance scheduling by modelling, among other things, inter-phase delay duration and time preference, in order to take advantage of shiftable loads and energy storage charging/discharging time. In [46], multi-microgrid configurations are presented and analyzed by means of the power line technology (AC, DC), layout (series, parallel, mixed), and interconnection technology (transformer, converter). A comparison of architectures based on cost, scalability, protection, reliability, stability, communications and business models is performed. Energy management and DR of multi-microgrids based on hierarchical multi-agent approach by introducing adjustable power is proposed by Bui et al. in [47]. Different operation modes are evaluated according to a two-level management cooperative multi-microgrid MILP-based model for day-ahead scheduling. Towards the application of state of the art, a microgrid energy management a Genetic Algorithm (GA) approach is applied in [48] to optimize cost strategies for scheduling distributed energy resources. The Quasi-static Artificial Bee Colony approach is used to optimize a multi-objective DR problem, based on the cost of energy and peak demand at the building level [49], including PV, Combined Heat and Power (CHP), batteries, electrical energy from the grid, and natural gas. Particle Swarm Optimisation is used in [50] to solve a bi-level problem modelling the interaction between the retailer and consumers. The energy hub is explored in [51] to develop a multi-carrier Demand-Side Management Time of Use (DSM ToU) optimization balancing energy import, conversion, and storage. Furthermore, a GA approach using present and day-ahead data was tested by Ferrari et al. [52] with respect to the management of loads of an experimental plant case study in Italy. The analysis involves PV, wind generation, a micro-CHP with a gas boiler, and an absorption chiller coupled with thermal storage.

In addition, Artificial Neural Networks (ANN)-based short term power forecasting is practiced to estimate day-ahead loads and renewable energy production. ANN models are designed to imitate biological nervous system information processing and evolution. They have been used for years in different areas of engineering, science, and business to deal with highly complex and nonlinear data sets. The ANN models assimilate the natural bonds of neurons and their high level interconnection to model complex systems. In the case of short-term predictions, the ANN models can be more effective compared to statistical, linear, or non-linear programming techniques. They encompass capabilities such as adaptive learning, self-organization, real time operation, fault tolerance, and the approximation of complex nonlinear functions. Kalaitzakis et al. in [53] tested advanced neural network short-term load forecasting using data from the electric power grid of the island of Crete in Greece. Various structures and configurations were assessed, and a parallel processing approach for a 24 h-ahead prediction was demonstrated. ANN architectures for forecasting demand in electric power systems are presented in [54] by Tsekouras et al. A case study of the Greek electric power grid is used to explore the performance of different ANN configurations and factors, including period length and inputs for training, confidence interval, and more. Moreover, short term power forecasting is of particular value for prosumers to model, understand, and predict their consumption profiles, as well as to apply effective scheduling and control. A framework for district-level energy management and ANN forecasting at the building level was investigated by Hu et al. in [55], evaluating the performance for 6 buildings of different occupancy routines. Hybrid Short Term Load Forecasting ANN combined with techniques such as Fuzzy Logic, GA, and Particle Swarm Optimisation are briefly discussed in [31]. Furthermore, a 24 h-ahead prediction of excess power at microgrid level is proposed by Mavrigiannaki et al. [56], testing 3 different configurations with respect to possible exploitation potentials from an energy management perspective. Finally, an overview of load forecasting, dynamic pricing, and demand side management techniques in smart grid research applications reveals the potential for operational cost reductions between 5–25% [57].

The aim of the research work presented here is the development and testing of a DR energy management GA-based optimisation approach based on day ahead ANN generated prediction models. The developed GA algorithm incorporates load shifting for the day ahead (24 h period), and evaluates possible alternatives based on cost and assumptions related to the practicality of the obtained solutions. The practical benefits of the proposed approach are linked to the development of a valuable tool for the evaluation of the potential rewards and risks of engagement in DR. In the case study that follows, a Time of Use pricing scheme is compared to a flat tariff.

The paper is organised as follows. In Section 2, the infrastructure and the applied methodology are presented. The proposed day-ahead GA approach for cost of energy and load shifting optimization based on ANN hourly power predictions is analyzed in Section 3. Results and considerations on ANN power predictions and GA optimisation solutions are provided in Section 4. Finally, in Section 5, conclusions and recommendations for future work are summarised.

2. Infrastructure and Methods

The proposed novel approach was developed and tested on the basis of data available from the MyLeaf platform, which monitors and controls the Leaf Community buildings. The Leaf Community is located in Angeli di Rosora, a small rural area in Marche region of Italy. It hosts industrial facilities of the Loccioni Group, a firm leading research and innovation activities in energy, environment, automotive, aviation, and other sectors. The Leaf community (Figure 1) consists of 5 industrial buildings (L3-AEA, L4-Leaf Lab, L5-Kite Lab, L6), one office building (L2-Summa), and a building used mainly for business meetings (Leaf Farm). All buildings (except the Leaf Farm) are equipped with rooftop photovoltaics (PV) of total power 629.2 kWp, and ground water heat pumps. In addition, a 2-axis solar tracker of 18 kWp, a 48 kWp micro-hydro plant, a 224 kWh battery storage, and a 523.25 kWh/K thermal storage are connected to the microgrid, which also features electric vehicle charging stations. Buildings, renewable energy systems (PVs, micro-hydro), and storage systems are all coupled and connected to the main power grid via a single interconnection line (point of delivery).

Figure 1. The Leaf Community map.

The buildings in the Leaf Community are highly thermally-insulated, and are equipped with automations for controlling the HVAC systems, as well as the natural and artificial lighting by means of adjustable external louvers and luminance sensors. The primary annual energy consumption for the Leaf Lab, is rated at 35.4 kWh/m^2 (including the PV power production and subtracting industrial consumption) [22], based on year-round measurements, while the L6 (new building, not fully operated yet), is estimated at 46.85 kWh/m^2. These ratings prove to be a factual determinant of their near zero energy performance. Table 1 summarizes the basic components of the building envelopes and systems installed at the Leaf Community buildings under consideration.

Table 1. Pilot buildings in the Leaf Community.

Pilot Case Studies	Sky Windows	Automatic Shading	Illuminance/ Presence Light Controls	LED	Ground Water Heat Pumps	biPV	Thermal Storage	Electrical Storage
Leaf Lab—Industrial (6000 m^2)	•	•	•	•	•	•	•	•
Summa—Offices/ Warehouse (1037 m^2)			•	•	•	•		•
Kite Lab (3514 m^2)— Offices, Laboratories	•		•	•	•	•		•

The elaborated methodology comprises several steps, as shown in Figure 2.

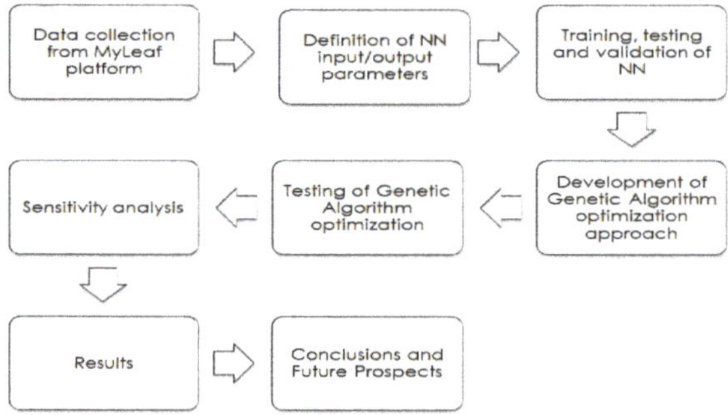

Figure 2. Methodological framework.

1. Collection of data: All data from measuring equipment, sensors, and actuators in the Leaf Community is collected, organized, and made remotely available through the MyLeaf platform [33]. In this case, the MyLeaf platform is used to collect data on the power demand of the buildings considered in the analysis.

2. Development and testing of ANN models: ANN models are developed and exploited to perform day-ahead predictions of consumption power using Matlab. For the 24 h-ahead prediction of power consumption, the day of week, the time, and the external temperature are used as inputs, while the 24 h-ahead electrical power is used as a target. Trials of various combinations for the ANN model parameterization are performed, considering the structure, algorithm, the number of hidden layers, and the delays. A Lavemberg-Marquardt algorithm was deployed in a Nonlinear Autoregressive ANN structure with Exogenous Input (NARX), with 3 hidden layers and a delay of 1.

3. GA approach: A genetic algorithm (GA) optimization scheme was developed and tested in Matlab, in order to provide alternative solutions for load shifting. The GA optimization scheme is based on the mathematical model analyzed in Section 3. The objective function encounters of the criteria of energy and load shifting. Market information is used to construct the hourly pricing profiles used in the optimization process. Weighting coefficients are applied to both normalized criteria to enable consideration of several alternatives, depending on several priorities, and energy management capabilities. Weighting coefficients are used to provide a trade-off between cost and load shift. The role of weighting coefficients is to allow a decision maker to investigate a set of solutions and obtain solutions which better match his/her preferences. Preferences differ based on the decision maker's knowledge and understanding, but may also be influenced by other factor priorities during the various time periods. For example, cost savings could be considered to be the "default" priority, but during certain periods, the minimization of load shifting could be upgraded to become the dominant factor in the optimization process.

4. Sensitivity analysis and evaluation of results: Sensitivity analysis is performed by changing the GA parameters, such as crossover, population size, mutation rate, tolerance etc. Furthermore, since load shifting is related to changes in the operation of building systems (HVAC, lighting, etc.) and operations (industrial, office), it also needs to be minimized in order to avoid significant intervention in the buildings' use. On the other hand, the cost of energy is minimized when load shifting occurs from hours of high prices to hours of low prices. The solutions are hence evaluated considering the hourly/daily cost of energy and load shifting preferences.

The developed approach is illustrated with the aid of the flowchart of Figure 3.

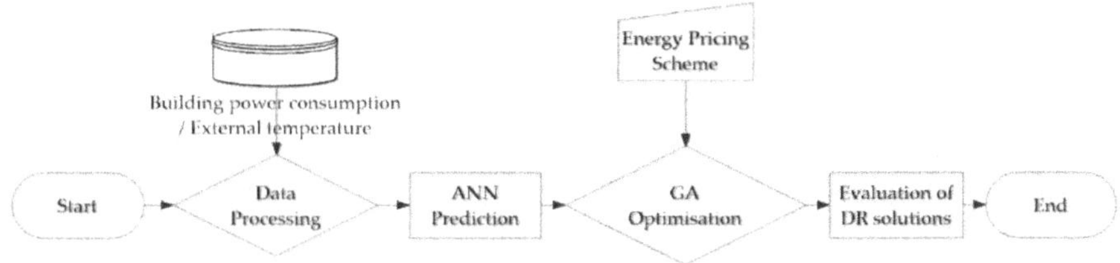

Figure 3. Flowchart of the developed approach.

3. The Proposed Day-Ahead GA Approach for Cost of Energy/Load Shifting Optimization Based on ANN Hourly Power Predictions

The GA optimisation scheme is based on the developed mathematical model presented hereafter. The two criteria, namely the normalised cost of energy and load shifting, form the objective function as shown in Equation (1):

$$f = \min\left(w_1 \frac{Cost_E}{Cost_{E_{max}}} + w_2 \frac{Load_{Shift}}{Load_{Shift\_max}} \right) \tag{1}$$

At building group level, the cost and load shift terms of the objective function in Equation (1), are given by Equations (2) and (6) which are further specified by Equations (3)–(5) and (7)–(9), respectively.

$$Cost_E = Cost_{E\_Lab} + Cost_{E\_Summa} + Cost_{E\_\_Kite} \tag{2}$$

Terms in Equation (2) are calculated based on Equations (3)–(5), as shown below:

$$Cost_{E\_Lab} = \sum_{h=1}^{24} X_{E\_\_Lab}^h * C_{E\_unit}^h \tag{3}$$

$$Cost_{E\_Summa} = \sum_{h=1}^{24} X_{E_{Summa}}^h * C_{E\_unit}^h \tag{4}$$

$$Cost_{E\_Kite} = \sum_{h=1}^{24} X_{E\_Kite}^h * C_{E_{unit}}^h \tag{5}$$

$$Load_{Shift} = Load_{Shift\_Lab} + Load_{Shift\_Summa} + Load_{Shift\_Kite} \tag{6}$$

where:

$$Load_{Shift\_Lab} = \sum_{h=1}^{24} abs(X_{E_{Lab}}^h - X_{E_{Lab_{baseline}}}^h) \tag{7}$$

$$Load_{Shift_{Summa}} = \sum_{h=1}^{24} abs\left(X_{E_{Summa}}^h - X_{E_{Summa_{baseline}}}^h \right) \tag{8}$$

$$Load_{Shift\_Kite} = \sum_{h=1}^{24} abs(X_{E_{Kite}}^h - X_{E_{Kite_{baseline}}}^h) \tag{9}$$

The following constraints in Equations (10)–(12) are applied to ensure that there is no deviation between the total daily energy consumed between baseline and the optimized solutions for each building:

$$\sum_{h=1}^{24} X_{E_{Lab}}^h - \sum_{h=1}^{24} X_{E_{Lab_{baseline}}}^h = 0 \tag{10}$$

$$\sum_{h=1}^{24} X_{E_{Summa}}^{h} - \sum_{h=1}^{24} X_{E_{Summa_{baseline}}}^{h} = 0 \tag{11}$$

$$\sum_{h=1}^{24} X_{E_{Kite}}^{h} - \sum_{h=1}^{24} X_{E_{Kite_{baseline}}}^{h} = 0 \tag{12}$$

Whether the optimization concerns a building or a building group analysis, for the evaluation of the GA based results, a comparison to baseline consumption, as obtained by the Artificial Neural Network day-ahead prediction, is conducted. The total cost linked to the genetic algorithm optimized solution is compared to the total cost of the baseline scenario, as evaluated by Equations (13) and (14) respectively:

$$Cost_{E\_opt} = \sum_{h=1}^{24} X_{E_{opt}}^{h} * C_{E_{unit}}^{h} \tag{13}$$

$$Cost_{E\_baseline} = \sum_{h=1}^{24} X_{E_{baseline}}^{h} * C_{E_{unit}}^{h} \tag{14}$$

4. Results and Discussion

4.1. ANN Based Predictions

The results of ANN-based predictions for the period from 1 May 2017 to 1 August 2017 and from 1 December 2017 to 1 March 2018, for each building, are presented in Figures 4 and 5, respectively. Day-ahead predicted values for Leaf Lab, Summa, and Kite Lab appear to be, in most cases, very close to real values, featuring a Pearson's correlation coefficient R in the range 0.96–0.98 for training, validation, testing, and overall. Lower R values are observed for Summa during the winter period.

At the left column of Figure 6, predicted versus real values of consumption power for the 3 buildings under study, are presented. At the right column of Figure 6, predicted versus real values of power are presented for the period from 12 February 2018 to 16 February 2018. Mean Bias Error (MBE) and Mean Average Percentage Error (MAPE) values, for the ANN predicted versus actual values on 21 July 2017 and 16 February 2018 for Leaf Lab, Summa, and Kite Lab, are presented in Table 2.

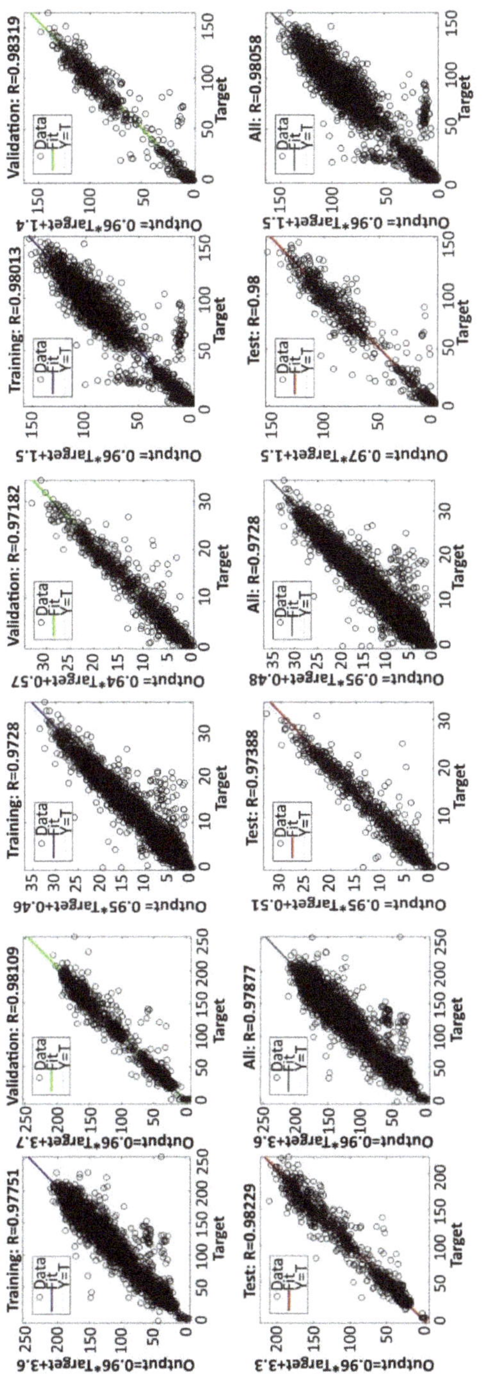

Figure 4. Prediction of electrical consumption power for Leaf Lab, Summa and Kite Lab from 1 May 2017 to 1 August 2017.

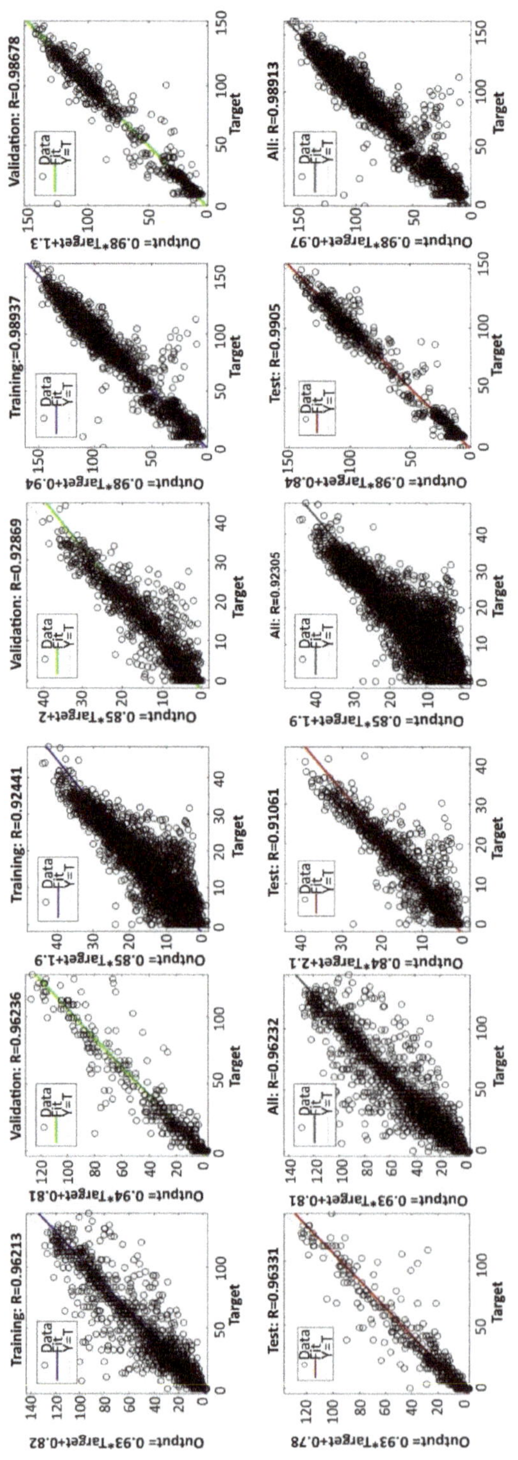

Figure 5. Prediction of electrical consumption power for the Leaf Lab, the Summa and the Kite Lab from 1 December 2017 to 1 March 2018.

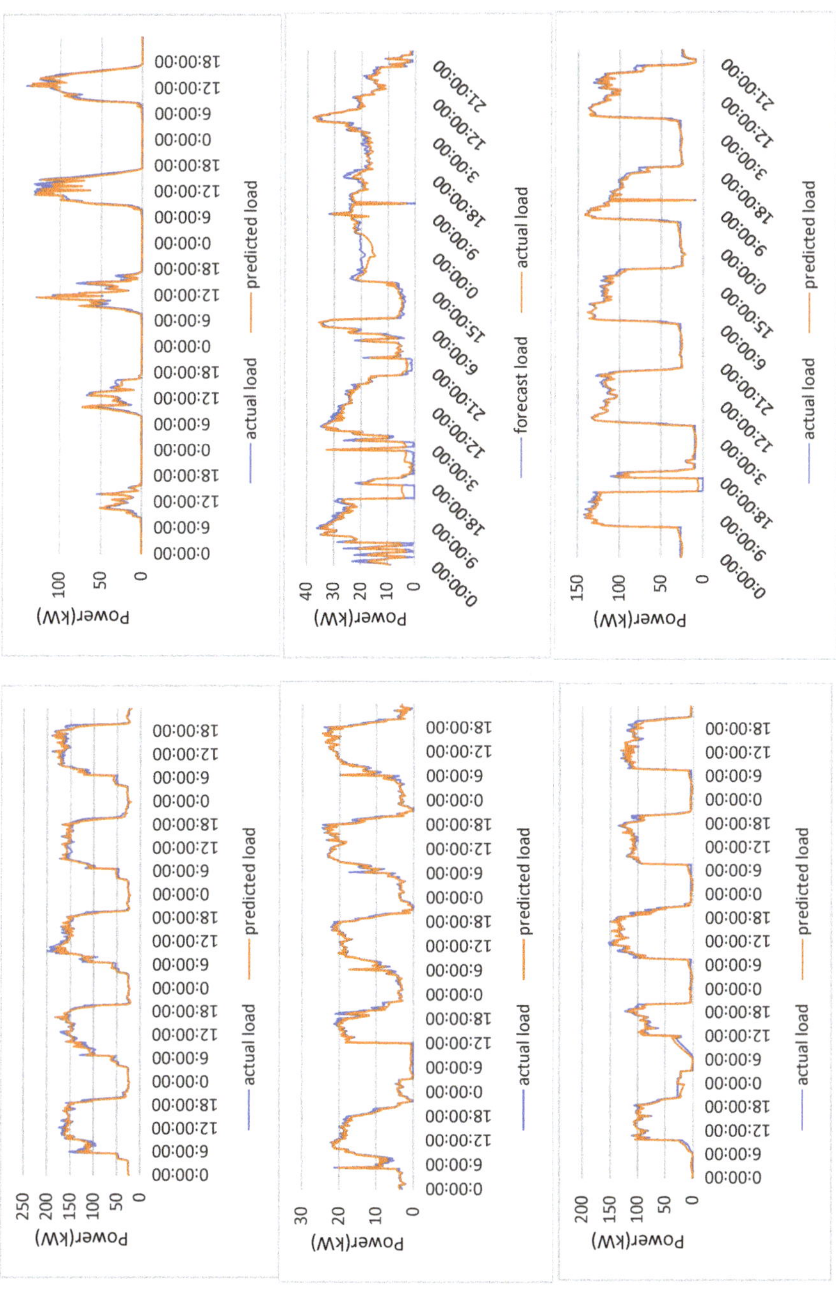

Figure 6. Prediction of electrical consumption power for the Leaf Lab, the Summa and the Kite Lab from 17 July 2017 to 21 July 2017 (**left**) and from 12 February 2018 to 16 February 2018 (**right**).

Table 2. MBE and MAPE for ANN predictions on 21 July 2017 and 16 February 2018.

ANN Prediction	21 July 2017		16 February 2018	
	MBE	MAPE (%)	MBE	MAPE (%)
Leaf Lab	1.43	5	−1.75	22.7
Summa	−0.01	8.47	−0.40	12
Kite Lab	−1.52	17.5	−1.42	4.96

4.2. Genetic Algorithm Optimization Results

In this section, the GA optimization results for 21 July 2017 and 16 February 2018 are presented and analyzed for the weighting coefficient values $w_1 = w_2 = 0.5$. For the baseline scenario, a flat tariff at 0.07 €/kWh is used. The optimized scenario is calculated taking into account a 2-zone tariff pricing scheme of 0.0675 €/kWh from 8 a.m. to 6 p.m., and 0.0525 €/kWh from 6 p.m. to 8 a.m. (Figure 7).

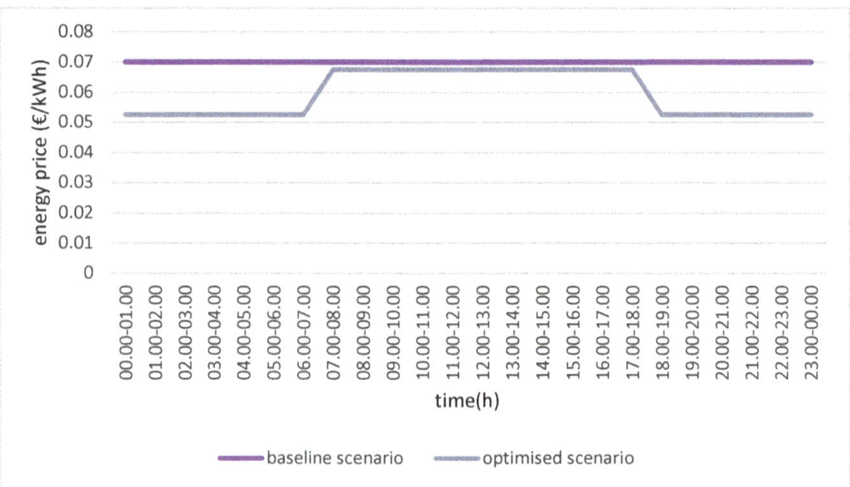

Figure 7. Energy pricing profiles used in the baseline and optimised scenarios.

In Figures 8 and 9, the results of the developed GA optimization approach are presented. The charts on the left columns of these figures illustrate the ANN-based power forecast as a baseline scenario. In the same charts, the GA optimized power profiles demonstrate load-shifting solutions. The related costs are depicted in the right columns of the Figures. The baseline costs are calculated based on the flat tariff of Figure 7, while the GA optimized costs are based on the 2-zone tariff of the same figure.

With respect to the Leaf Lab, it is observed in Figure 8 that load shifting occurs from the high-price to low-price hours. This is also reflected, in terms of the cost profile, on the day which accounts for a reduction of 15.77% from €174.97 to €147.37. Likewise, the load in Summa is shifted outside the high price region, with the baseline daily cost of €20.55 being decreased down to €17.80, a relevant reduction of 13.38%. Similarly, the load shifting in Kite Lab occurs from the high tariff zone towards the morning and the evening hours, without a reduction in total power consumption. In this case, the baseline cost is €101.89, and the optimized total cost is €87.40, achieving a reduction percentage of 14.22%.

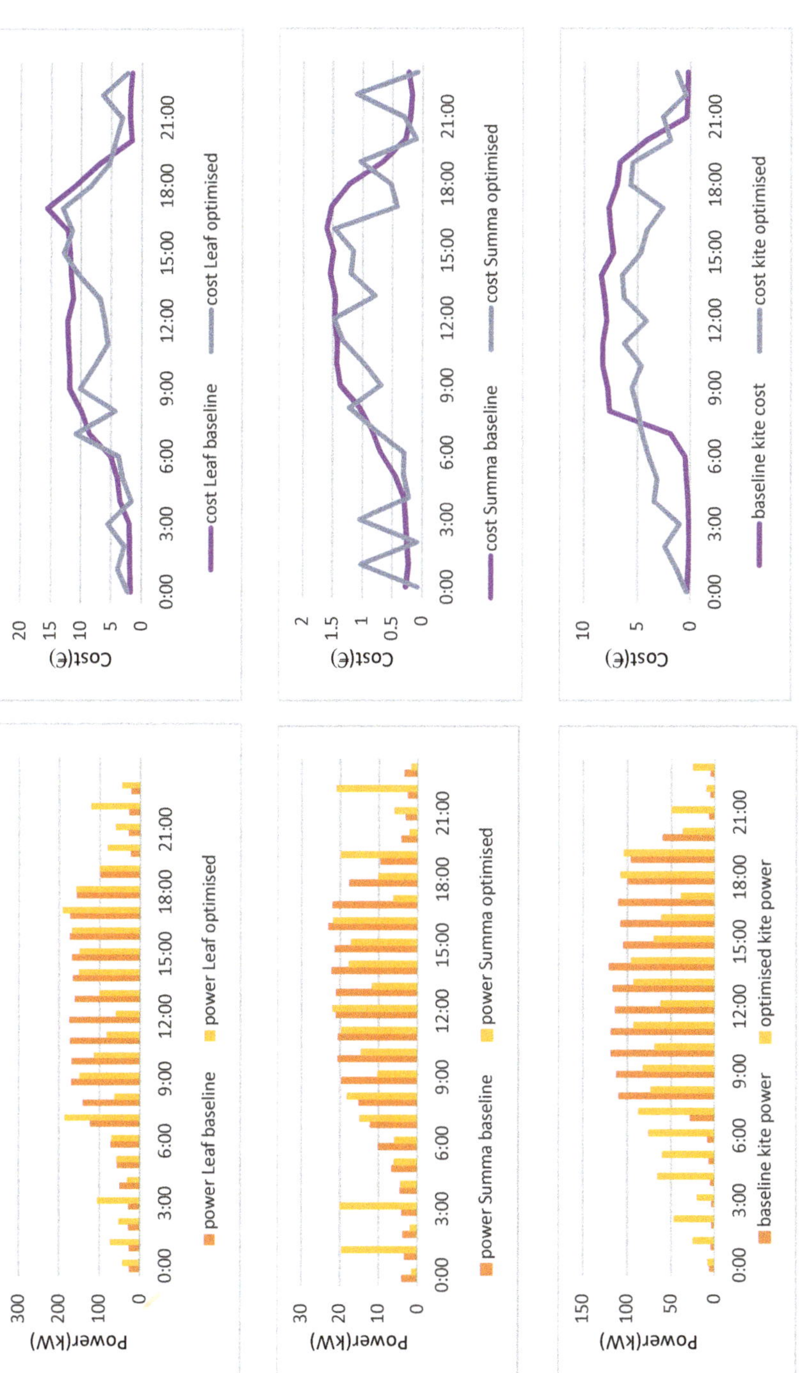

Figure 8. GA optimisation power and cost results for the Leaf Lab, the Summa and the Kite Lab on 21 July 2017.

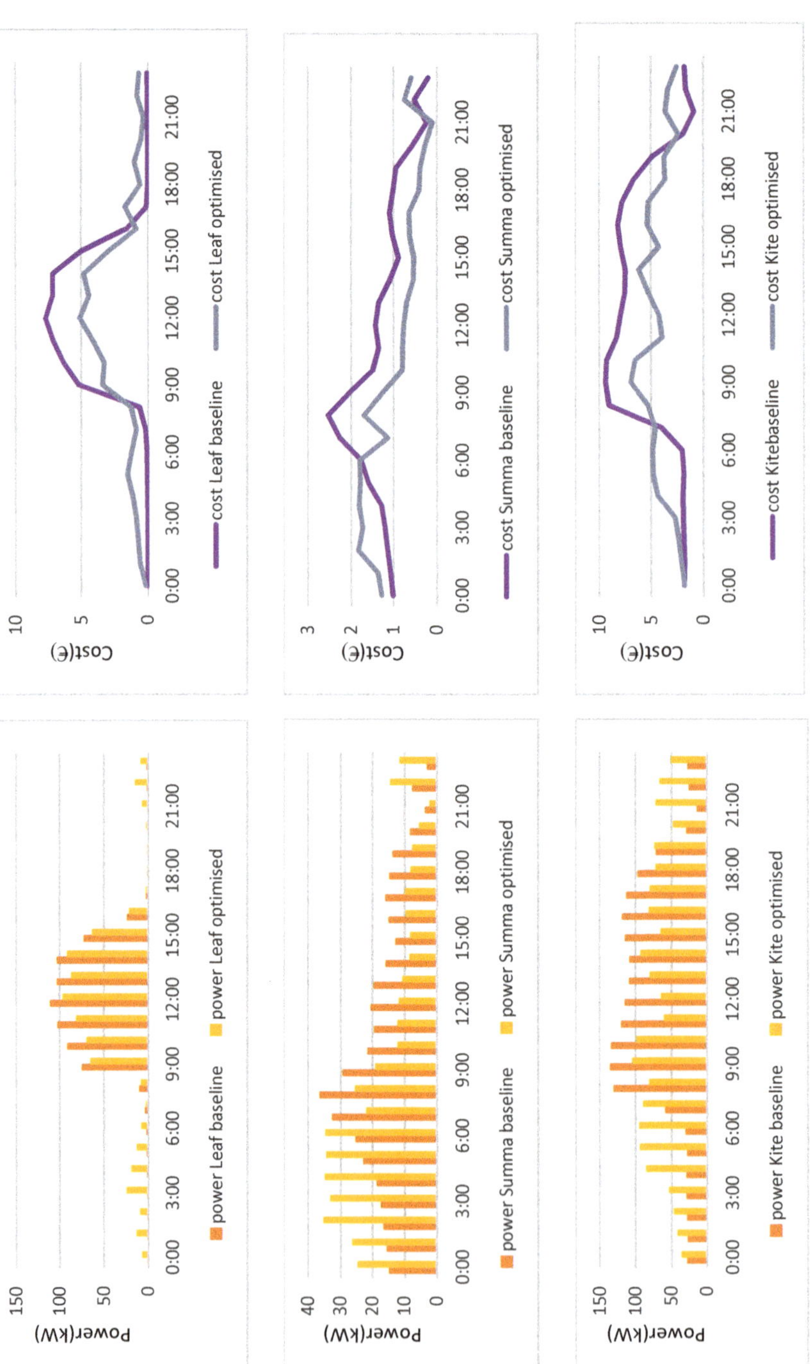

Figure 9. GA optimisation power and cost results for the Leaf Lab, the Summa and the Kite Lab during 16 February 2018.

The analysis of the winter results is displayed in Figure 9. The shift for the Leaf Lab power profile leads to a cost reduction of 10.23% from €48.21 in the baseline scenario down to €43.27.

Load shifting throughout the 24 h occurs in Summa in a way that changes the overall power profile to the early hours of the day. This transition of loads corresponds to 18.59% of costs savings, reflecting also the differences between the flat and the 2-zone tariff pricing scheme.

With respect to the daily power in Kite Lab during the winter, changes between baseline and optimized scenarios appear to take place in a harmonic way from high to low price hours. In this case, a 14.97% cost saving is achieved, since the baseline daily cost is €118.1, compared to the optimized daily cost of €100.42.

In Figure 10, the total power consumption of the 3 buildings is illustrated. In the first case, the high power consumption according to the baseline power is shifted from working hours towards early morning and late evening hours. In terms of cost, the total baseline cost at district level is 293.95 €, and the total optimized cost is €251.30, corresponding to a 14.51% reduction.

With respect to the winter period, the hourly-district level GA-optimized power values for equal weighting coefficients undergo a smooth differentiation to the left of the graph with respect to the baseline. The district-level total baseline cost is €195.27, and the total optimized cost is €167, achieving a reduction percentage of 14.47%.

According to Table 3, regarding the Leaf Lab, the results for each case prove that the optimization is successful, bearing in mind that the baseline cost is €174.90 and the optimized values range from €142.24 to €153.97, a maximum operational costs percentage reduction of 18.67%. For Summa, as for the Leaf Lab, the optimized cost for each pair of weights is lower than the baseline cost of €20, and varies between €17.80 and €17.12. The percentage reduction in this case reaches 14.4% Furthermore, the optimization for the Kite Lab revealed that the GA produces better results compared to the baseline cost of €101.9 for all pairs of weights ranging from €87.62 down to €85.78. The percentage reduction in this case is up to 15.82%. The last column of the table represents the optimized cost for the group of buildings, which is lower than the baseline cost of €293 for all pairs of weighting coefficients varying from €253.91 to €247.89. The maximum percentage reduction in this case is 15.39%.

Table 3. Results of the optimization on 21 July 2017 during the summer period.

w_1	w_2	Leaf Lab Cost (€)	Summa Cost (€)	Kite Lab Cost (€)	District Level Cost (€)
0	1	153.97	17.40	86.48	252.09
0.1	0.9	149.80	17.467	87.07	250.76
0.2	0.8	152.15	17.742	86.76	253.91
0.3	0.7	145.71	17.517	87.60	251.91
0.4	0.6	148.44	17.21	87.34	253.70
0.5	0.5	147.37	17.80	87.46	251.30
0.6	0.4	151.51	17.784	87.62	252.24
0.7	0.3	152.21	17.39	87.23	251.38
0.8	0.2	149.69	17.457	86.92	247.89
0.9	0.1	144.40	17.466	85.78	251.77
1	0	142.24	17.12	86.62	251.78

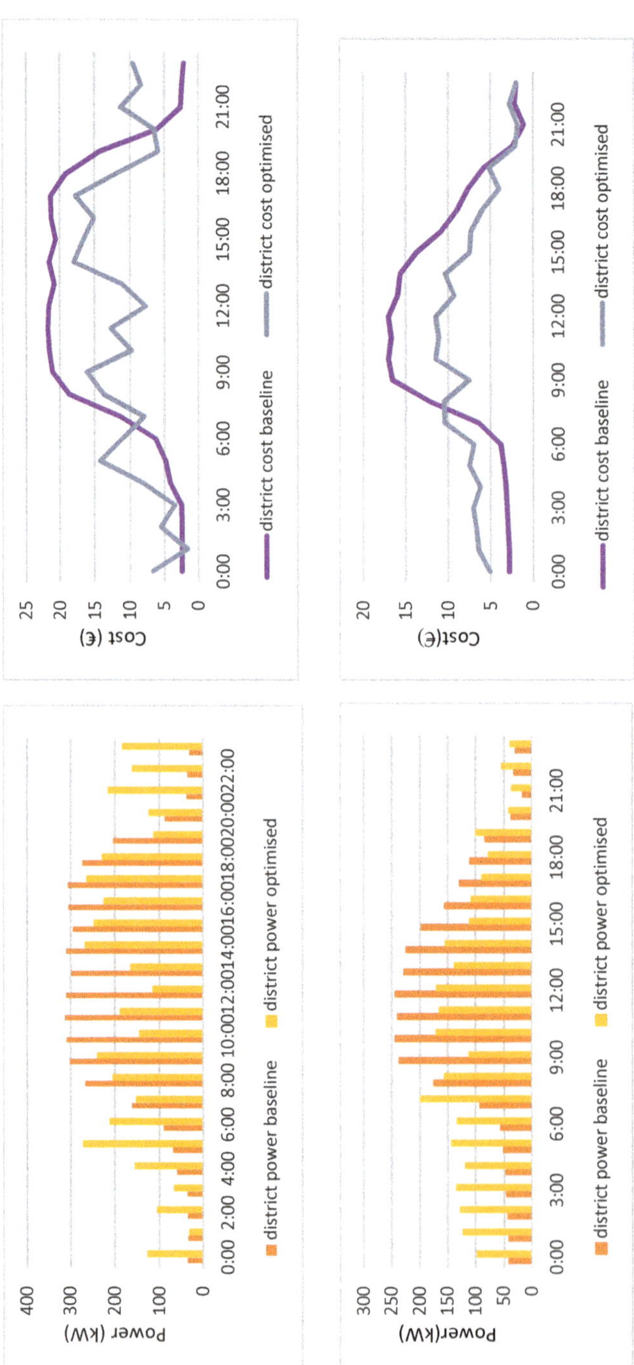

Figure 10. GA optimisation power and cost results for the total power on 21 July 2017 (**up**) and 16 February 2018 (**down**).

Table 4 includes the results of optimization for each pair of weighting coefficients at both the building and district levels for the winter period. The results for the Leaf Lab depict the optimized cost for all weights combinations. As shown, in all cases, the optimized cost varies between €43.48 to €42.81, which is lower than the baseline cost of €48 in this case, and accounts for a percentage reduction of up to 10.81%. Moreover, the optimization for the Summa building revealed genetic algorithm solutions with costs from €23.58 to €23.27, a percentage reduction of 16.89% compared to the baseline cost of €28 in this building. Subsequently, in the Kite Lab, the optimized cost is from €102.06 down to €99.85, equal to a percentage reduction of up to 15.38% lower than the baseline cost of €118. The last column represents the optimized cost in the group of buildings during the winter, varying from €168.19 to €166.53, leading to a percentage reduction of 14.6% compared to the baseline cost of €195.

Table 4. Results of the optimization on 16 February 2018 during the winter period.

w_1	w_2	Leaf Lab Cost (€)	Summa Cost (€)	Kite Lab Cost (€)	District Level Cost (€)
0	1	42.87	23.34	101.04	167.40
0.1	0.9	42.81	23.36	101.04	167.33
0.2	0.8	42.94	23.43	99.85	167.97
0.3	0.7	43.48	23.40	101.19	167.66
0.4	0.6	43.18	23.56	102.06	168.19
0.5	0.5	43.28	23.58	100.42	167.01
0.6	0.4	43.05	23.52	101.25	166.53
0.7	0.3	43.29	23.39	101.81	167.94
0.8	0.2	43.07	23.33	100.43	167.45
0.9	0.1	43.04	23.49	100.49	167.43
1	0	42.92	23.27	101.07	166.67

4.3. Limitations of the Adopted Two-Level Model

The proposed approach entails some level of abstraction with respect to the load shift which is achievable within the capacity of individual systems and components. Evaluating load shift in conjunction with a pricing scheme requires deep knowledge, and depends on the specificities of each case study. In this respect, load shift is determined by technical factors, i.e., installed systems technical characteristics, control scheme, etc., as well as organisational factors, i.e., the potential shift of the industrial operations within each building. A detailed knowledge of the operation of each system in a building, along with data, i.e., power consumption profile, is not available in most cases. This logic can be applied to some extent by using constraints to ensure that a specific percentage of the power at any time remains unchanged. Consequently, optimisation can be conducted based on the flexible share of the consumption power for every hour.

Also, the proposed approach is linked to the accuracy of the prediction, which may vary according to the building under study, and other factors, e.g., type of loads, industrial operations, season, etc. Therefore, it is important to evaluate the risk associated with different prediction error levels according to the examined pricing scheme. Although this risk is low in a two zone pricing scheme, it may become significant when considering dynamic pricing profiles.

5. Conclusions

The main contribution of this work is related to linking ANN short-term electric forecasting and GA multi-objective optimization as a tool for generating and evaluating alternative day-ahead load shifting solutions. The first step of the proposed approach is exploiting Artificial Neural Network modelling for the prediction of the power consumption in a period of 24 h ahead. Predictions of hourly-consumption power levels using day of week, time of day, and external temperature as inputs were obtained for each of the 3 buildings of the Leaf Community (Leaf Lab, Summa, and Kite Lab). The results proved that a close correlation between predicted and actual values exists during the studied summer and winter periods, as evaluated based on correlation coefficient R for the whole period, as well as Mean Bias Error (MBE) and Mean Average Predicted Error (MAPE) for the specific days used in the optimization process.

The second step was to create an optimization function to include energy cost and load shifting using appropriate variables and constraints. The objective function was minimized using a Genetic Algorithm to obtain solutions at individual building and building group levels. Results demonstrated the effectiveness of this approach in considering alternative pricing schemes and load shifting possibilities as a way to examine cost savings. Cost savings of between 10.81% and 18.67% at the building level were associated with significant load shifting solutions obtained by the GA scheme in the considered two-zone ToU pricing scheme. At the district level, cost savings in the range of 13.34% and 15.39% were obtained.

Future steps in this work may involve: (i) extending research activities to include renewable energy generation and storage capabilities, (ii) reforming the GA obtained solutions so as to take into consideration actual loads (base, fixed, flexible), renewable energy production, and storage, and (iii) exploiting the potential for improvements in power predictions using ANN models.

Author Contributions: Conceptualization, N.K. and D.K.; methodology, N.K., E.T. and D.K.; software, N.K. and E.T.; validation, N.K., E.T. and D.K.; formal analysis, N.K.; investigation, N.K., D.K. and E.T.; resources, D.K., C.C. and D.I.; data curation, N.K. and E.T.; writing—original draft preparation, N.K. and E.T.; writing—review and editing, N.K., E.T., D.K. and K.K.; visualization, N.K., D.K. and E.T.; supervision, D.K. and K.K.; project administration, D.K. and C.C.; funding acquisition, D.K., N.K. and C.C.

Funding: This project has received funding from the European Union's Horizon 2020 research and innovation programme under the Marie Skłodowska-Curie grant agreement No 645677.

Conflicts of Interest: The authors declare no conflict of interest.

Abbreviations

AC	Alternated Current
ADR	Automated Demand Response
AMI	Advanced Metering Infrastructure
ANN	Artificial Neural Network
ARC	Aggregators or Retail Customers
BEMs	Building Energy Management Systems
CHP	Cogeneration of Heat and Power
CPP	Critical Peak Pricing
CSP	Curtailment Service Providers
DC	Direct Current
DEMs	District Energy Management Systems
DER	Distributed Energy Resources
DR	Demand Response
DRP	Demand Response Provider
DSM	Demand Side Management
EED	Energy Efficiency Directive
GA	Genetic Algorithm
HVAC	Heating, Ventilation, Air Conditioning
HRES	Hybrid Renewable Energy Systems
IoT	Internet of Things
MINLP	Mixed Integer Non Linear Programming
PV	Photovoltaic
PSO	Particle Swarm Optimisation
RTP	Real Time Pricing
ToU	Time of Use

Nomenclature

$Cost_E$	district daily energy operating costs (€)
$Cost_{E_{max}}$	normalisation factor of cost criterion (€)
$Cost_{E\_Lab}$	daily energy operating costs of Leaf Lab (L4) building (€)
$Cost_{E\_Summa}$	daily energy operating costs of Summa (L2) building (€)
$Cost_{E\_Kite}$	daily energy operating costs of Kite (L5) building (€)
$C^h_{E\_unit}$	day-ahead hourly unit cost of energy in each building (€/kWh)
$Cost_{E\_baseline}$	total cost of the baseline scenario
$Cost_{E\_opt}$	total cost of the genetic algorithm optimised solution
$Load_{Shift}$	daily load shift (kWh)
$Load_{Shift\_max}$	normalisation factor of load shift criterion (kWh)
$Load_{Shift\_Lab}$	daily load shift of Leaf Lab (L4) building (kWh)
$Load_{Shift\_Summa}$	daily load shift of Summa (L2) building (kWh)
$Load_{Shift\_Kite}$	daily load shift of Kite (L5) building (kWh)
w_1	Weighting coefficient of cost criterion [0–1]
w_2	Weighting coefficient of load shift criterion [0–1]
X^h_E	hourly value of total energy consumption in each building (kWh)
$X^h_{E_{Lab}}$	hourly value of total energy consumption in Leaf Lab (L4) building (kWh)
$X^h_{E_{Lab_{baseline}}}$	Baseline (predicted) hourly value of total energy consumption in Leaf Lab (L4) building (kWh)
$X^h_{E_{Summa}}$	hourly value of total energy consumption in Summa (L2) building (kWh)
$X^h_{E_{Summa_{baseline}}}$	baseline (predicted) hourly value of total energy consumption in Summa (L2) building (kWh)
$X^h_{E\_Kite}$	hourly value of total energy consumption in Kite (L5) building (kWh)
$X^h_{E_{Kite_{baseline}}}$	baseline (predicted) hourly value of total energy consumption in Kite (L5) building (kWh)
$X^h_{E_{opt}}$	GA optimised hourly electrical energy (kWh) at building or building group level
$X^h_{E_{baseline}}$	baseline hourly electrical energy (kWh) based on day-ahead Neural Network predictions

References

1. Greer, C.; Wollman, D.A.; Prochaska, D.E.; Boynton, P.A.; Mazer, J.A.; Nguyen, C.T.; FitzPatrick, G.J.; Nelson, T.L.; Koepke, G.H.; Hefner, A.R., Jr.; et al. *NIST Framework and Roadmap for Smart Grid Interoperability Standards, Release 3.0*; Special Publication (NIST SP): Gaithersburg, MD, USA, 2014.
2. Rietveld, G.; Clarkson, P.; Wright, P.S.; Pogliano, U.; Braun, J.-P.; Kokalj, M.; Lapuh, R.; Zisky, N. Measurement infrastructure for observing and controlling smart electrical grids. In Proceedings of the 2012 3rd IEEE PES International Conference and Exhibition on Innovative Smart Grid Technologies (ISGT Europe), Berlin, Germany, 14–17 October 2012.
3. NERC. *Demand Response Availability Data System (DADS): Phase I & II Final Report*; North American Electric Reliability Corporation: Atlanta, GA, USA, 2011.
4. European Parliament. *Directive 2012/27/EU of the European Parliament and of the Council of 25 October 2012 on Energy Efficiency*; European Parliament: Brussels, Belgium, 2012; pp. 1–56.
5. Bertoldi, P.; Zancanella, P.; Boza-Kiss, B. *Demand Response Status in EU Member States*; EUR 27998 EN; Joint Research Centre: Ispra, Italy, 2016.
6. OpenADR Alliance. *OpenADR 2.0 Demand Response Program Implementation Guide*; OpenADR Alliance: San Ramon, CA, USA, 2014.
7. Siano, P. Demand response and smart grids—A survey. *Renew. Sustain. Energy Rev.* **2014**, *30*, 461–478. [CrossRef]
8. Aalami, H.A.; Moghaddam, M.P.; Yousefi, G.R. Demand response modeling considering Interruptible/Curtailable loads and capacity market programs. *Appl. Energy* **2010**, *87*, 243–250. [CrossRef]
9. Kusiak, A.; Tang, F.; Xu, G. Multi-objective optimization of HVAC system with an evolutionary computation algorithm. *Energy* **2011**, *36*, 2440–2449. [CrossRef]
10. Ghazvini, M.A.F.; Soares, J.; Morais, H.; Castro, R.; Vale, Z. Dynamic pricing for demand response considering market price uncertainty. *Energies* **2017**, *10*, 1–20.

11. Triki, C.; Violi, A. Dynamic pricing of electricity in retail markets. *4OR* **2009**, *7*, 21–36. [CrossRef]
12. Haider, H.T.; See, O.H.; Elmenreich, W. Residential demand response scheme based on adaptive consumption level pricing. *Energy* **2016**, *113*, 301–308. [CrossRef]
13. Tsui, K.M.; Chan, S.C. Demand Response Optimization for Smart Home Scheduling under Real-Time Pricing. *IEEE Trans. Smart Grid* **2012**, *3*, 1812–1821. [CrossRef]
14. Crosbie, T.; Broderick, J.; Short, M.; Charlesworth, R.; Dawood, M. Demand Response Technology Readiness Levels for Energy Management in Blocks of Buildings. *Buildings* **2018**, *8*, 13. [CrossRef]
15. Crosbie, T.; Short, M.; Dawood, M.; Charlesworth, R. Demand response in blocks of buildings: Opportunities and requirements. *Entrep. Sustain. Issues* **2017**, *4*, 271–281. [CrossRef]
16. Attia, M.; Haidar, N.; Senouci, S.M.; Aglzim, E.H. Towards an efficient energy management to reduce CO2emissions and billing cost in smart buildings. In Proceedings of the 2018 15th IEEE Annual Consumer Communications & Networking Conference (CCNC), Las Vegas, NV, USA, 12–15 January 2018; pp. 1–6.
17. Kolokotsa, D. The role of smart grids in the building sector. *Energy Build.* **2016**, *116*, 703–708. [CrossRef]
18. Karlessi, T. The Concept of Smart and NZEB Buildings and the Integrated Design Approach. *Procedia Eng.* **2017**, *180*, 1316–1325. [CrossRef]
19. Diakaki, C.; Grigoroudis, E.; Kolokotsa, D. Towards a multi-objective optimization approach for improving energy efficiency in buildings. *Energy Build.* **2008**, *40*, 1747–1754. [CrossRef]
20. Kolokotsa, D.; Diakaki, C.; Grigoroudis, E.; Stavrakakis, G.; Kalaitzakis, K. Decision support methodologies on the energy efficiency and energy management in buildings. *Adv. Build. Energy Res.* **2009**, *3*, 121–146. [CrossRef]
21. Diakaki, C.; Grigoroudis, E.; Kabelis, N.; Kolokotsa, D.; Kalaitzakis, K.; Stavrakakis, G. A multi-objective decision model for the improvement of energy efficiency in buildings. *Energy* **2010**, *35*, 5483–5496. [CrossRef]
22. Kampelis, N.; Gobakis, K.; Vagias, V.; Kolokotsa, D.; Standardi, L.; Isidori, D.; Cristalli, C.; Montagnino, F.M.; Paredes, F.; Muratore, P.; et al. Evaluation of the performance gap in industrial, residential & tertiary near-Zero energy buildings. *Energy Build.* **2017**, *148*, 58–73.
23. Gobakis, K.; Mavrigiannaki, A.; Kalaitzakis, K.; Kolokotsa, D.-D. Design and development of a Web based GIS platform for zero energy settlements monitoring. *Energy Procedia* **2017**, *134*, 48–60. [CrossRef]
24. Kolokotsa, D.; Gobakis, K.; Papantoniou, S.; Georgatou, C.; Kampelis, N.; Kalaitzakis, K.; Vasilakopoulou, K.; Santamouris, M. Development of a web based energy management system for University Campuses: The CAMP-IT platform. *Energy Build.* **2016**, *123*, 119–135. [CrossRef]
25. Zhou, B.; Li, W.; Chan, K.W.; Cao, Y.; Kuang, Y.; Liu, X.; Wang, X. Smart home energy management systems: Concept, configurations, and scheduling strategies. *Renew. Sustain. Energy Rev.* **2016**, *61*, 30–40. [CrossRef]
26. Shakeri, M.; Shayestegan, M.; Abunima, H.; Reza, S.M.S.; Akhtaruzzaman, M.; Alamoud, A.R.M.; Sopian, K.; Amin, N. An intelligent system architecture in home energy management systems (HEMS) for efficient demand response in smart grid. *Energy Build.* **2017**, *138*, 154–164. [CrossRef]
27. Fanti, M.P.; Mangini, A.M.; Roccotelli, M.; Ukovich, W.; Pizzuti, S. A Control Strategy for District Energy Management. In Proceedings of the 2015 IEEE International Conference on Automation Science and Engineering (CASE), Gothenburg, Sweden, 24–28 August 2015; pp. 432–437.
28. Marinakis, V.; Doukas, H. An advanced IoT-based system for intelligent energy management in buildings. *Sensors (Switzerland)* **2018**, *18*, 610. [CrossRef] [PubMed]
29. Brusco, G.; Burgio, A.; Menniti, D.; Pinnarelli, A.; Sorrentino, N.; Scarcello, L. An Energy Box in a Cloud-Based Architecture for Autonomous Demand Response of Prosumers and Prosumages. *Electronics* **2017**, *6*, 98. [CrossRef]
30. Lu, N.; Du, P.; Guo, X.; Greitzer, F.L. Smart meter data analysis. In Proceedings of the Pes T&D, Orlando, FL, USA, 7–10 May 2012; pp. 1–6.
31. Baliyan, A.; Gaurav, K.; Mishra, S.K. A review of short term load forecasting using artificial neural network models. *Procedia Comput. Sci.* **2015**, *48*, 121–125. [CrossRef]
32. Raza, M.Q.; Khosravi, A. A review on artificial intelligence based load demand forecasting techniques for smart grid and buildings. *Renew. Sustain. Energy Rev.* **2015**, *50*, 1352–1372. [CrossRef]
33. Shinkhede, S.; Joshi, S.K. Hybrid Optimization Algorithm for Distributed Energy Resource. In Proceedings of the 2014 Innovative Applications of Computational Intelligence on Power, Energy and Controls with their impact on Humanity (CIPECH), Ghaziabad, India, 28–29 November 2014.

34. Goulden, M.; Bedwell, B.; Rennick-Egglestone, S.; Rodden, T.; Spence, A. Smart grids, smart users? the role of the user in demand side management. *Energy Res. Soc. Sci.* **2014**, *2*, 21–29. [CrossRef]

35. Guo, S.; Liu, Q.; Sun, J.; Jin, H. A review on the utilization of hybrid renewable energy. *Renew. Sustain. Energy Rev.* **2018**, *91*, 1121–1147. [CrossRef]

36. Dawoud, S.M.; Lin, X.; Okba, M.I. Hybrid renewable microgrid optimization techniques: A review. *Renew. Sustain. Energy Rev.* **2018**, *82*, 2039–2052. [CrossRef]

37. Ou, T.-C.; Lu, K.-H.; Huang, C.-J. Improvement of Transient Stability in a Hybrid Power Multi-System Using a Designed NIDC (Novel Intelligent Damping Controller). *Energies* **2017**, *10*, 488. [CrossRef]

38. Ou, T.-C.; Hong, C.-M. Dynamic operation and control of microgrid hybrid power systems. *Energy* **2014**, *66*, 314–323. [CrossRef]

39. Hatziargyriou, N.; Asano, H.; Iravani, R.; Marnay, C. Microgrids. *IEEE Power Energy Mag.* **2007**, *5*, 78–94. [CrossRef]

40. Zia, M.F.; Elbouchikhi, E.; Benbouzid, M. Microgrids energy management systems: A critical review on methods, solutions, and prospects. *Appl. Energy* **2018**, *222*, 1033–1055. [CrossRef]

41. Bui, V.-H.; Hussain, A.; Kim, H.-M.; Bui, V.-H.; Hussain, A.; Kim, H.-M. Optimal Operation of Microgrids Considering Auto-Configuration Function Using Multiagent System. *Energies* **2017**, *10*, 1484. [CrossRef]

42. Ou, T.-C. A novel unsymmetrical faults analysis for microgrid distribution systems. *Int. J. Electr. Power Energy Syst.* **2012**, *43*, 1017–1024. [CrossRef]

43. Ou, T.-C. Ground fault current analysis with a direct building algorithm for microgrid distribution. *Int. J. Electr. Power Energy Syst.* **2013**, *53*, 867–875. [CrossRef]

44. Hirsch, A.; Parag, Y.; Guerrero, J. Microgrids: A review of technologies, key drivers, and outstanding issues. *Renew. Sustain. Energy Rev.* **2018**, *90*, 402–411. [CrossRef]

45. Mohseni, A.; Mortazavi, S.S.; Ghasemi, A.; Nahavandi, A.; Abdi, M.T. The application of household appliances' flexibility by set of sequential uninterruptible energy phases model in the day-ahead planning of a residential microgrid. *Energy* **2017**, *139*, 315–328. [CrossRef]

46. Bullich-Massagué, E.; Díaz-González, F.; Aragüés-Peñalba, M.; Girbau-Llistuella, F.; Olivella-Rosell, P.; Sumper, A. Microgrid clustering architectures. *Appl. Energy* **2018**, *212*, 340–361. [CrossRef]

47. Bui, V.-H.; Hussain, A.; Kim, H.-M. A Multiagent-Based Hierarchical Energy Management Strategy for Multi-Microgrids Considering Adjustable Power and Demand Response. *IEEE Trans. Smart Grid* **2018**, *9*, 1323–1333. [CrossRef]

48. Wang, Y.; Huang, Y.; Wang, Y.; Zeng, M.; Li, F.; Wang, Y.; Zhang, Y. Energy management of smart micro-grid with response loads and distributed generation considering demand response. *J. Clean. Prod.* **2018**, *197*, 1069–1083. [CrossRef]

49. Safamehr, H.; Rahimi-Kian, A. A cost-efficient and reliable energy management of a micro-grid using intelligent demand-response program. *Energy* **2015**, *91*, 283–293. [CrossRef]

50. Carrasqueira, P.; Alves, M.J.; Antunes, C.H. Bi-level particle swarm optimization and evolutionary algorithm approaches for residential demand response with different user profiles. *Inf. Sci.* **2017**, *418–419*, 405–420. [CrossRef]

51. Batić, M.; Tomašević, N.; Beccuti, G.; Demiray, T.; Vraneš, S. Combined energy hub optimisation and demand side management for buildings. *Energy Build.* **2016**, *127*, 229–241. [CrossRef]

52. Ferrari, L.; Esposito, F.; Becciani, M.; Ferrara, G.; Magnani, S.; Andreini, M.; Bellissima, A.; Cantù, M.; Petretto, G.; Pentolini, M. Development of an optimization algorithm for the energy management of an industrial Smart User. *Appl. Energy* **2017**, *208*, 1468–1486. [CrossRef]

53. Kalaitzakis, K.; Stavrakakis, G.S.; Anagnostakis, E.M. Short-term load forecasting based on artificial neural networks parallel implementation. *Electr. Power Syst. Res.* **2002**, *63*, 185–196. [CrossRef]

54. Tsekouras, G.J.; Kanellos, F.D.; Mastorakis, N. Chapter 2 Short Term Load Forecasting in Electric Power Systems with Artificial Neural Networks. In *Computational Problems in Science and Engineering*; Springer International Publishing: Cham, Switzerland, 2015; Volume 343.

55. Hu, J.; Leung, V.C.M.; Coulson, G.; Ferrari, D. *Smart Grid Inspired Future Technologies*; Springer: Berlin, Germany, 2017; Volume 203.

56. Mavrigiannaki, A.; Kampelis, N.; Kolokotsa, D.; Marchegiani, D.; Standardi, L.; Isidori, D.; Christalli, C. Development and testing of a micro-grid excess power production forecasting algorithms. *Energy Procedia* **2017**, *134*, 654–663. [CrossRef]

57. Khan, A.R.; Mahmood, A.; Safdar, A.; Khan, Z.A.; Khan, N.A. Load forecasting, dynamic pricing and DSM in smart grid: A review. *Renew. Sustain. Energy Rev.* **2016**, *54*, 1311–1322. [CrossRef]

Review

Energy Resources, Load Coverage of the Electricity System and Environmental Consequences of the Energy Sources Operation in the Slovak Republic—An Overview

Martin Lieskovský [1], Marek Trenčiansky [1], Andrea Majlingová [2,*] and Július Jankovský [3]

[1] Faculty of Forestry, Technical University in Zvolen, T. G. Masaryka 24, 96001 Zvolen, Slovakia; martin.lieskovsky@tuzvo.sk (M.L.); marek.trenciansky@tuzvo.sk (M.T.)

[2] Faculty of Wood Science and Technology, Technical University in Zvolen, T. G. Masaryka 24, 96001 Zvolen, Slovakia

[3] Apertis, s.r.o. (LLC), V. P. Tótha 17, 960 01 Zvolen, Slovakia; julius.jankovsky@apertis.eu

* Correspondence: majlingova@tuzvo.sk; Tel.: +421-45-520-6065

Received: 1 March 2019; Accepted: 30 April 2019; Published: 6 May 2019

Abstract: According to the current circumstances that are related to the effectiveness of the tightened European Union (EU) environmental legislation, which sets minimum requirements for the heat and power sources of energy that are part of the Slovak Electricity System (SES) source mix, an article was prepared to summarize the information regarding energy and environmental legislation, which is in force as in the EU as in Slovakia. This information was completed with a description on the current situation and requirements for the safety and reliability of the "new" mix of sources and technologies of electricity production within the SES in terms of energy and economic efficiency and environmental consequences.

Keywords: energy resource; energy security; energy sources; Slovakia

---

1. Introduction

The start of electricity usage reaches the end of the 19th century. First, electricity has begun to be used for lighting, and later as a source of power for mechanical drives. Further, its use has gradually spread to other sectors, too. Today, electricity has penetrated all areas of the economy and the social spheres. The production and consumption of electricity places demands on the reliability of the elements of the power system and the creation of capacity reserves at the same time. Energy needs to have enough primary energy resources and the availability of flexible technologies. Whereas, in the past, the regulatory reserves have been dimensioned and activated to compensate for the deviations in the consumption or for bridging the failures of production capacities. Resource reserves must now also be activated in balancing the variations in the change in the volume of electricity production to sources with unpredictable direct production in the direct conversion of solar and wind energy. Such production, due to the preferential connection with the power system, requires back-up with predictive sources in the full potential of its installed power.

The availability of electricity is taken for granted and the break-up of the system is not allowed in the European Union (EU). Electricity has become available to subscribers, of course, so almost nobody is interested in the technical nature of its production and distribution. The consequence of this interpretation is the misconception that it is necessary to pay "exorbitant" high bills for an ordinary socket and switch. The electricity bill in the family budget, even in the corporate budget, accounts for a fraction of the cost of mobile operator services, the Internet, licenses for software, hardware, car leasing, and other substantially less important items that would be "non-functional scrap" without electricity.

Overall, it is forgotten that electricity in the socket and switch is continuously available thanks to the reliable operation of the power grid, which is the most important, the most extensive, and most expensive infrastructure in each EU country, while the transmission system connections are subjected to the Union for the Co-ordination of Transmission of Electricity (UCTE) in the EU and North Africa, the rules of which must be strictly observed by national operators both dispatching and commercial dealers. The life of a civilized society without electricity is impossible. Of course, the consequences of a power failure are sometimes perceived during natural disasters quite differently to the ones that are directly affected by the failure. Power-supply engineers warn that reliable operation is not a matter of course. Since the beginning of the millennium, nine large disruptions of the power systems ("blackouts") had occurred in the world, and even the smallest had impact on more inhabitants, as it lives in the whole Slovakia. Overall, more than a billion inhabitants of the planet have been affected by the disruption of the power system since the beginning of this century. Five of the all disruptions have taken place in the EU in the last 18 years.

The largest electricity failure in the EU in this century occurred at the end of September 2003. It initiates a series of network failures in northern Italy, which were caused by the fall of a tree on the electric power transmission line in Switzerland during a storm. Within four seconds, the failure spread to Italy, Slovenia, and Croatia. The disruption of the electricity power system has affected 55 million people; fortunately it only lasted for a short time, one day. Life in Italy, except Sardinia, has stopped. Rail and air travel completely collapsed, cars were moving sharply across the streets while the traffic lights were not operating, and thousands of them stood without fuel at the edge of the road due to the non-operating filling stations. The mass media and the internet ceased to exist. The public did not know what had happened. Banks, ATMs, shops, restaurants, and offices were totally collapsing. The lives of people in hospitals saved electricity from backup sources. Fortunately, the disintegration of the electric system occurred at the end of summer at 2.00 pm at night. If the failure would occur in the winter, during a cold weather, the consequences would be catastrophic when the outside temperature can fall below the freezing point in the temperate climatic zone.

2. European and Slovak Legislation Governing the Energy Sector

In January 2007, the European Commission (EC) published the Communication "An Energy Policy for Europe" [1]. This Communication outlined the developments in the energy sector by 2010, as well as the 2020 targets. This Communication, while respecting the sovereignty and energy mix of the individual EU countries, integrated energy policy with climate change policy and clearly formulated the three basic pillars of EU energy policy that were: energy security; competitiveness; and, sustainability.

Subsequently, in March 2007, the European Council adopted the Energy Efficiency Action Plan 2007–2010 [2], which is an important element of climate change commitments: reducing greenhouse gas emissions by 20% by 2020 as compared to 1990; increasing the share of renewable energy to 20% by 2020; achieve a 10% share of renewable energy in transport by 2020; and, achieve 20% energy savings as compared to projections by 2020.

The Energy Efficiency Action Plan has become the basic document for the development of the legislative framework in the upcoming period. It was followed by the other strategic and legislative documents that covered the different areas of the Action Plan: Strategic Energy Technology Plan (2007) [3], Third Liberalization Package (2007) [4], Climate-Energy Package (2008) [5] and Energy Efficiency Plan (2006–2011) [6]. The Second Strategic Energy Review—An EU Energy Security and Solidarity Action Plan (2008) [7] focused on the least developed energy policy pillar—energy security, just in gas crisis time in January 2009. The European Economic Recovery Plan (2008) [8] included a proposal to support the development of energy infrastructure, with the support of specific projects in the area of gas infrastructure development in the Slovak Republic.

An important milestone in the development of energy policy was the adoption of the Treaty of Lisbon [9] in 2009. The Treaty on the functioning of the EU defined a new legal basis for EU energy policy measures and its Article. 194 defined the basic objectives and principles of EU energy policy. The main objectives of the European energy policy are to ensure the functioning of the energy market; ensuring security of energy supply in the EU; enhancing energy efficiency and energy savings and developing new technologies for electricity production and promoting the production of electricity from renewable energy sources, as well as supporting the interconnection of energy systems and networks. The sovereignty of the Member States in the composition of the energy mix as well as in ensuring their energy security are enshrined in the basic principles of European energy policy.

The 2020 energy principles and targets were based on the Europe 2020 Strategy [10] and they are further elaborated in the Energy 2020 strategy: A Strategy for a Competitive, Sustainable, and Secure Energy [11]. The key energy priorities included: to make effective use of energy resources in the EU, to complete a pan-European integrated energy market by 2015, to increase the rights of consumers and to achieve an increase in energy security, to maintain the EU's leading role in energy technology, and to strengthen the external dimension of the EU energy market.

The energy efficiency demands have been gradually becoming a centre of interest, as evidenced by the revision of the EU's energy efficiency policy in the form of the adoption of the Energy Efficiency Directive 2012/27/EU [12]. This Directive established a common framework for measures to promote energy efficiency in the EU in order to secure the EU's main energy efficiency target of a 20% reduction in energy consumption by 2020 under the Europe 2020 Strategy [10].

In the field of energy infrastructure, in November 2010, the Communication "Energy Infrastructure Priorities for 2020 and Beyond" [13] identified the key roles for the needs of infrastructure development in the oil, gas, and electricity sectors by 2020 and the basic long-term and short-term (by 2020) priorities in the field of European energy infrastructure that are needed to complete the interconnection of the internal market. These include the North-South Gas and Electricity Interconnections, the oil connections in Central Europe, and the Southern Gas Corridor, regarding the development of energy infrastructure in Central Eastern and South-Eastern Europe, with relevance to the Slovak Republic. These priority corridors of the European Energy Infrastructure were further elaborated in the "Energy Infrastructure Package" that was proposed in 2011 and adopted in 2013 by the European Parliament and Council (EP and C) no. 347/2013 on the Guidelines for Trans-European Energy Infrastructure (TEN-E) [14] and Regulation 1316/2013 [15] establishing the Connecting Europe Facility (CEF). The decision-making body adopted in July 2013 a European list of projects of common interest in the electricity, gas, and oil sectors.

The EC in the Roadmap to a Competitive Low Carbon Economy in 2050 (03/2011) [16] analysed the implications of a commitment to reduce greenhouse gas emissions by 80–95% by 1990 and indicated the extent of emission reductions in the key sectors for the years 2030 and 2050. Electricity will play a central role in the low-carbon economy. The Commission's analysis shows that, by 2050, it can contribute to almost complete elimination of CO_2 emissions and offers prospective partial replacement of fossil fuels in transport and heating. The EC calls on the other European institutions and the Member States to take this plan into account in the further development of European, national, and regional policies that are aimed at building a low carbon economy by 2050.

In the Energy Roadmap 2050 (12/2011) [16], in several scenarios, the EC is exploring ways of decarbonizing the energy system and ways of securing energy supply and competitiveness by 2050. The plan seeks to develop a long-term, technologically neutral European framework for energy policies, thereby achieving the necessary certainty and stability in investing in the energy system. The Roadmap does not replace the national, regional, and local efforts to modernize energy supply, it but seeks to develop a long-term, technologically neutral, European framework, in which these policies will be more effective.

The EC published the Communication "Renewable Energy: A Major Player in the European Energy Market" (06/2012) [17], which is aimed at ensuring the growth of sustainability beyond 2020. The Communication set out key priorities, such as increased coordination of support systems, strengthening the role of the Southern Mediterranean, the use of cooperative mechanisms, and progress in the field of energy technologies.

In March 2013, the EC issued the Green Paper: A 2030 Framework for Climate Change Policy and Energy Policies [18], and it launched a debate on the Energy and Climate Framework Policy Post-2020 at the same time. On 22 January 2014, the EC published the EC's Communication on an EU policy: 2030 Climate and Energy Framework [19], following the Green Paper from March 2013. In March 2014, EC accepted the commitment to adopt the 2030 Climate and Energy Framework policy by October 2014. The Slovak Republic still does not have a definitive position in this respect, although the coordination of positions takes place between the ministries concerned (Ministry of Finance SR, Ministry of Environment SR, and Ministry of Economy SR). During the negotiations on the future framework, the Ministry of Economy emphasized the need to preserve sovereignty in the field of energy mix, the non-binding character of the 2020 renewable energy sources (RES) and energy efficiency targets, the need to respect the national specifics, and the need to develop RES in a cost-effective way. It is a binding target for reducing greenhouse gas emissions by 2030, which was subjected to certain conditions.

The EC identified in the Communication "Progress towards Completing the Internal Energy Market" [20] (11/2012), the obstacles, and measures that are needed to meet the objective of completing the EU internal energy market (IEM) by 2014 and removing the isolation of Member States by the year 2015. The Communication also contained recommendations for the Slovak Republic concerning the elimination of the regulation of energy supply prices, the solution of round-the-clock issues, and the development of North-South connections in the gas and electricity sectors.

At the end of 2018, the EC concluded negotiations on all aspects of the new energy legislative framework "Clean Energy for All Europeans package" [21]. This is a significant step towards the creation of the Energy Union and delivering on the EU's Paris Agreement commitments. The package includes eight different legislative acts: "Energy Performance in Buildings Directive"; "Renewable Energy Directive"; "Energy Efficiency Directive"; "Governance Regulation"; "Electricity Directive"; "Electricity Regulation"; "Risk-Preparedness Regulation"; and, "Regulation for the Agency for the Cooperation of Energy Regulators (ACER)". This new policy framework brings regulatory certainty, through the introduction of the first national energy and climate plans. It will encourage essential investments to take place in the energy sector. It empowers European consumers to become fully active players in the energy transition and it fixes two new targets for the EU for 2030: a binding renewable energy target of at least 32% and an energy efficiency target of at least 32.5%—with a possible upward revision in 2023. For the electricity market, it confirms the 2030 interconnection target of 15%, following on from the 10% target for 2020. When these policies are fully implemented, they will lead to steeper emission reductions for the whole EU than anticipated—some 45% by 2030 relative to 1990 (as compared to the existing target of a 40% reduction).

That legislative framework also played an important part in the EC's preparations for its long-term strategy for a climate neutral Europe by 2050 ("Climate Neutral Economy by 2050" [22]), which was published in November 2018. The strategy shows how Europe can lead the way to climate neutrality by investing into realistic technological solutions, empowering citizens, and aligning action in key areas, such as industrial policy, finance, or research—while ensuring social fairness for a just transition. It is in line with the Paris Agreement objective to keep the global temperature increase to well below 2 °C and to pursue efforts to keep it to 1.5 °C. The purpose of this long-term strategy is not to set targets, but to create a vision and sense of direction, plan for it, and inspire as well as enable stakeholders, researchers, entrepreneurs, and citizens alike to develop new and innovative industries, businesses, and associated jobs. It investigates the portfolio of options that are available for Member States, business, and citizens, and how these can contribute to the modernisation of our economy and improve the quality of life of Europeans. The long-term strategy also seeks to ensure that this transition is socially fair and it enhances the competitiveness of EU economy and industry on global markets, securing high quality jobs and sustainable growth in Europe, while also helping to address other environmental challenges, such as air quality or biodiversity loss. The road to a climate neutral economy would require joint action in seven strategic areas: energy efficiency; deployment of renewables; clean, safe and connected mobility; competitive industry and circular economy, infrastructure, and interconnections; bio-economy and natural carbon sinks; and, carbon capture and storage to address remaining emissions.

The energy legislative framework in the Slovak Republic is based on several documents that can be divided into three groups in general. The first group consists of primary legislation. There are the acts that were adopted by the National Council of the Slovak Republic. The second group of documents (secondary legislation) consists of generally binding legal regulations (decrees) of the Ministry of Economy of the Slovak Republic, the Regulatory Office for Network Industries (URSO), or the Slovak Government. The last group of legislative documents is tertiary legislation, which includes the operating rules, technical conditions for access, and connection to the system and networks, dispatching orders and documents that are drawn up and published in accordance with the provisions of Act no. 251/2012 Coll. on Energy [23] and Act no. 250/2012 Coll. on Regulation in Network Industries [24].

The primary electricity legislation is represented by Act no. 251/2012 Coll. on Energy, Act no. 250/2012 Coll. on Regulation in Network Industries, and Act no. 476/2008 Coll. on Energy Efficiency.

The EU's third energy package was implemented in 2009 by issuing Act no. 250/2012 Coll. [25]. The Act provides greater independence and power for the regulatory authority to determine regulated prices as well as control activities in regulated entities. By implementing the EU's third liberalization package in the Slovak legislation, space for reducing the regulatory burden in the energy sector has been opened for the future. The Act in question regulates: the subject, scope, conditions, and method of regulation in network industries; the conditions for carrying out regulated activities and the rights and obligations of regulated entities; the rules for the functioning of the electricity and gas markets; establishment, status, and competence of the Regulatory Office for Network Industries (URSO) and the Regulatory Council; proceedings under this Act; and, administrative offenses for breach of obligations that are specified by this Act. The subject of regulation under this Act is the determination or approval of the method, procedures, and conditions for: connection and access to the transmission system, distribution system, transmission network and distribution network; electricity transmission and distribution in a defined area; gas transmission and gas distribution in a defined area; providing support services in the electricity and gas industries; providing transmission system operator and distribution system operator services, accessing and connecting new electricity and gas producers to the system or network, producing and distributing heat.

Act no. 251/2012 Coll. on Energy [23] is the basic legislative document of the Slovak energy sector. The Act regulates: the conditions for the functioning of the open energy market; the rights and obligations of individual electricity market participants and gas market participants; network and systems management issues, i.e., power dispatching and gas dispatching centres; issues of separation of regulated activities, i.e., unbundling; special forms of electricity production; network access conditions; rules for market behaviour of energy market participants; and, performance of state administration and supervision in the energy sector. From 1 July 2007, all customers, including households, have the option of choosing their energy supplier (electricity and gas) based on the open market.

Act no. 476/2008 Coll. on Energy Efficiency [25] sets out: the concept and action plans for energy use; evaluation of transmission, transport and distribution; energy consumption obligations in buildings; an obligation on the manufacturer to operate, reconstruct and build energy-efficient energy conversion equipment; for a manufacturer to carry out an energy audit to demonstrate the possibility of supplying usable heat; and, consumers of energy in industry and in agriculture the obligation to evaluate energy intensity of production by energy audits, conditions of operation of the monitoring system.

Another important act from the energy point of view is the Act no. 309/2009 Coll. on the Promotion of Renewable Energy Sources and High Efficiency Cogeneration and on Amendments to Certain Acts [26]. This Act specifies: the method of support and conditions for the promotion of electricity production from renewable energy sources, electricity by high-efficiency cogeneration, biomethane; rights and obligations of producers of electricity from renewable energy sources, electricity from cogeneration, electricity from high-efficiency cogeneration, biomethane; the rights and obligations of other electricity and gas market participants; and, the rights and obligations of the legal person or the natural person who places on the market fuels and other energy products used for transport purposes.

The secondary legislation is primarily represented by the Decree of the Office for Regulation of Network Industries no. 24/2013 Coll., laying down the rules for the functioning of the domestic market in electricity and the rules for the functioning of the domestic gas market [27] (including the Decree of the Office for Regulation of Network Industries no. 423/2013 Coll.).

This Decree lays down detailed rules for the functioning of the electricity and gas markets when the electricity market participant is connected to the system, the electricity market participant's access to the system, electricity transmission, cross-border electricity exchange, electricity distribution, electricity supply, including supply of regulation electricity and electricity supply to households, providing support services, providing system services, assuming responsibility for deviation, evaluating, clearing and settlement of electricity market participant's deviation and system deviations, and how to prevent system overloading and system overload solutions. It also discusses the procedure for changing the electricity supplier, the conditions, and the date of its implementation in detail.

Other important regulations are:

- Decree of the Ministry of Economy no. 599/2009 Coll. aimed at implementing the provisions mentioned in the Act on the Promotion of Renewable Energy Sources and High Efficiency Cogeneration. [28]
- Decree of the Office for Regulation of Network Industries no. 80/2015 Coll. amending the Decree of the Regulatory Office for Network Industries no. 490/2009 Coll. laying down details on the promotion of renewable energy sources, high efficiency cogeneration and biomethane. [29]
- Decree of the Office for Regulation of Network Industries no. 189/2014 Coll. amending the Decree of the Regulatory Office for Network Industries no. 221/2013 Coll. laying down the price regulation in the electricity industry. [30]

From January 2019 an amendment to the Act no. 309/2009 Coll. on the Promotion of Renewable Energy Sources and High-Efficiency Cogeneration and on Amendments to Certain Acts is in force [26]. In connection with this amendment, the Act No. 251/2012 Coll. on Energy was also amended. The biggest changes concern the payment for access to the distribution system (so-called G-component). This payment was introduced with effect from January 2014 by the Decree of the Office for Regulation of Network Industries focusing prices. In 2016, the Constitutional Court annulled the G-component and declared that part of the Decree imposing an obligation on the G-components to pay electricity producers to be unconstitutional. According to the Constitutional Court, the legal basis for the payment of the G-component can only be the agreement on access to the distribution system and the distribution of electricity. The amendment of this Act responds to this decision in several points. The first change is the modification of access to the distribution system definition. Access to the distribution system also means the supply of electricity to the system. The essential requirements of the agreement regarding access to the distribution system and electricity distribution are also changed. If the contract is concluded by the electricity producer, the transmission of electricity is not an essential requirement of the contract. Another change is the introduction of a definition of unauthorized supply. Unauthorized supply is the supply of electricity to the system without a contract on access to the distribution system and the distribution of electricity. If the producer delivers electricity without an access agreement, it will be an unauthorized delivery and the distribution company will be able to disconnect it. The last important change is the obligation for electricity producers to conclude a contract on access to the distribution system and electricity distribution system to with a system operator when they supply electricity to the system.

Act no. 250/2012 Coll. on the Regulation in Network Industries was also amended. The amendment introduces a statutory definition of tariffs for system operation and tariffs for system services. So far, these terms have only been governed by legal regulations. At the same time, there is a legal definition of one of the components of the tariff for operating the system, which is the tariff to produce electricity from renewable energy sources. There are also individual tariffs for operating the system for businesses with a consumption of at least 1GWh (energy-intensive businesses). The extended competences of the Office for Regulation of Network Industries to issue a price decision for RES and CHP electricity generation activities for the entire period of support by a supplement or surcharge, i.e., for a period of 15 years, are introduced. The amendment also ensures the effective implementation of the decisions of the Agency for the Cooperation of Energy Regulators (ACER). As a rule, ACER decisions have the character of methodologies and conditions. The amendment introduces an obligation for the Office for Regulation of Network Industries to publish these decisions on its website in order to ensure that electricity market participants and gas market participants are properly informed regarding their issue. It also provides for the possibility for the Office for Regulation of Network Industries to stop the price proceedings, if, by issuing a price decision, the total installed capacity of the new electricity generation facilities from RES and CHP would be exceeded. The Ministry of Economy of the SR for the relevant calendar year will publish the total installed capacity of new equipment.

3. Basic Information about the Slovak Electricity System

On the territory of Slovakia, the beginning of the electrification of the territory relates to the operation of the first hydroelectric power plants in the Central and Northern Slovakia at the end of the 19th century, which gradually merged into local systems. The first brown-coal-fired steam power plant, with 12.0 MW installed capacity, was put into operation in Handlova more than 100 years ago and became the primary source of the regional system in the area between Handlova and Prievidza. In 1949, a brown-coal-fired power station in Novaky with an installed capacity of 178 MW was commissioned. In 1963, a new 110 MW unit was put into operation in Novaky, connected to the Bystricany substation, and then connected to the already highly developed Czech electricity system through the 110-kV substation Liskovec in the north of Moravia. The operation of the first system power plant guaranteed the security of electricity supplies to Slovakia, which, until then, had been mainly supplied by electricity from a hydroelectric Vah river cascade which reliability was limited by the potential of the Vah's hydro energy. At present, the electrical system is an infrastructure property of high value with thousands of resources, tens of thousands of kilometres of superior and distribution air and cable lines, appliances, and safety features of the power grid. The electrical system facilities have made electricity continuously available throughout the country, and it serves the development of the economy and the population of the Slovak Republic.

It is important that a balance of production and consumption of electricity is always maintained in order to guarantee the safety and reliability of the electrical system operation. The frequency and voltage offset reflects the difference in electricity production and consumption, while the higher the power production and lower the power consumption, the frequency increases, and vice-versa. If the frequency (49.8–50.2 Hz) and voltage offset is exceeded, their life span would be reduced and their own protections would start to be disconnected from the system. According to the current load, the difference in MW production and consumption may cause local outages in the supply of electricity, which in the undesired process ends with the disruption of the system. Disturbances in the system spread at a speed that goes beyond the possibilities of human perception. Therefore, each regulatory area must have a system of protection and management of the electricity system that prevents the spread of faults and controls the production of resources and consumption of consumers in the nodes, while real controlling the operation of the power system of the Slovak Republic (SR) is provided by the "Slovenska elektrizacna prenosova sustava" (SEPS Inc., Zilina)/Slovak Electricity Transmission System.

4. Production, Distribution and Consumption of Electricity in Slovakia

A power system of 7721 MW was installed in 2017 and the system integrated the production of electricity from almost 2800 electricity production sources, according to the Yearbook 2017 of the SEPS [31]. Electricity consumption reached 31,066 GWh, of which the domestic production was accounted for 28,036 GWh. The electricity production balance reached 3 030 GWh, i.e., almost 10%. The annual power maximum of 4550 MW was reached on January 11, 2017, and the power minimum of 2380 MW on 21 May 2017, the regulatory range of the system was of 2170 MW. The share of production from fossil fuels, nuclear sources, and renewable energy sources (RES), as well as the import of electricity, has been stabilized for the long time, and optimal in terms of safety and reliability. Regulatory requirements increase due to the unpredictable change in the power of photovoltaic power plants, with an installed power potential of 540 MW. When compared to 2015, the installed capacity was reduced by more than 400 MW. Four units of coal power plants were decapitated, of which two 110 MW blocks belonged to the Coal Power Plant Novaky and two blocks of equal power belonged to the Coal Power Plant Vojany. The reason for the decapitation was the failure to comply with the prescribed environmental requirements of the EU and SR. The Ministry of Economy of the Slovak Republic did not allow further decapitation of system resources installed capacity in order to maintain energy security of Slovakia, unless they are put into operation the domestic sources, i.e., new blocks of the nuclear power plant in Mochovce.

Electricity that is produced in the Slovak Republic is loaded with low greenhouse gas emissions. In 2017, only 20% of electricity was produced from fossil fuels. The total load of electricity that was consumed by CO_2 emissions was 230 g/kWh, which is the seventh lowest position within the 28 EU countries.

Table 1 shows the share of resources for installed power and total electricity production, costs, and prices of electricity produced and imported. Table 1 provides a detailed overview of the major electricity producers (about 90% of electricity production in Slovakia) with the volume of emitted CO_2, NO_X, and SO_X.

Table 1. Production and consumption of electricity in Slovakia in 2017 (Source: [31,32]).

Energy Sources	Installed Power (MW)	(%)	Electricity Production 2017 (GWh)	(%)	Use (%)	Price (€/MWh)	Costs (K €)
Nuclear	1940.0	25,1	15,081.0	53.8	88.7	36.20	545,932
Water	2539.0	32.9	4667.0	16.6	21.0	45.71	213,329
Lignite	333.0	4.3	1734.0	6.2	59.4	97.44	168,968
Hard coal	221.0	2.9	1062.0	3.8	54.9	86.93	92,324
Natural gas	1106.0	14.3	2228.0	7.9	23.0	77.72	173,169
Oil	257.0	3.3	687.0	2.5	30.5	100.60	69,115
Mixed fuels	431.0	5.6	90.0	0.3	2.4	91.93	8274
Sun	530.0	6.9	609.0	2.2	13.1	385.73	234,912
Biomass	225.0	2.9	1113.6	4.0	56.5	97.59	108,681
Biogas	105.0	1.4	642.0	2.3	69.8	134.51	86,358
Wind	3.0	0.0	3.9	0.0	14.8	64.59	252
Others RES	12.0	0.2	76.5	0.3	72.8	113.11	8653
Other	19.0	0.2	42.0	0.1	25.2	26.93	1131
Production	7721.0	100.0	28,036.0	100.0	41.5	61.03	1,711,098
Import/Export	345.9	4.5	3030.0	9.8	100.0	36.20	109,686
Consumption	8066.9	104.5	31,066.0	100.0	44.0	58.61	1,820,784

Among the "other RES" belongs, e.g., geothermal energy or sewage sludge processing in Slovakia.

The price of electricity from photovoltaics is several times higher than from coal due to the high investment and costs and lower efficiency of utilization of this RES. For solar energy, the average use of installed power is only 13.1% and for coal 55–60%. Coal from Slovak mines is subsidized by the state, so the price is as high as the price of biomass.

The price of electricity and natural gas is rising, mainly due to the dramatic increase in the price of greenhouse gas emission allowances. Paradoxically, in Slovakia at the end of 2018, an Act was issued to stop the combined production of electricity and heat from biomass. Electricity that is produced in this way is considered to be particularly environmentally friendly in a civilized world, because, in addition to high energy efficiency, it slows down climate change, as it is neutral in terms of greenhouse gas production. Moreover, the prices of heat that is produced in the cogeneration process together with electricity would not increase for more than 400 thousand of the inhabitants of Slovakia, as the purchase of emission allowances does not affect this production.

Figure 1 introduces the end user price of electricity according to the energy sources.

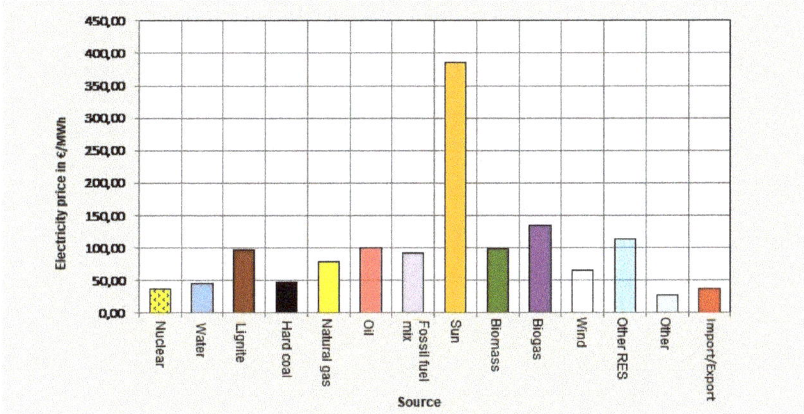

Figure 1. Comparison of commodities end user prices (Source: [31,32]).

Figure 2 shows the electricity production and consumption in Slovakia, as well as the electricity production and consumption balance.

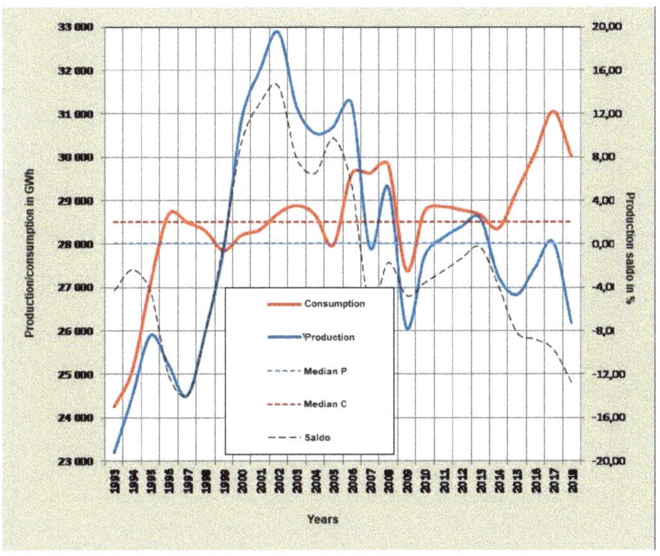

Figure 2. Production and consumption of electricity in Slovakia (Source: [31,32]).

Figure 2 shows the decrease in electricity production. The reason is the termination of the operation of the first block V-1 of nuclear power plant in Jaslovske Bohunice in January 2007 as well as its second block V-1 in January 2009. Each of these V-1 blocks generated 10% of all energy that is produced in Slovakia. The country is expected to become self-sufficient again after putting the third and fourth block of the nuclear power plant in Mochovce into operation (2019, 2020).

Figures 3–5 show the commodities end user prices for electricity, gas, and EUA's emission allowances.

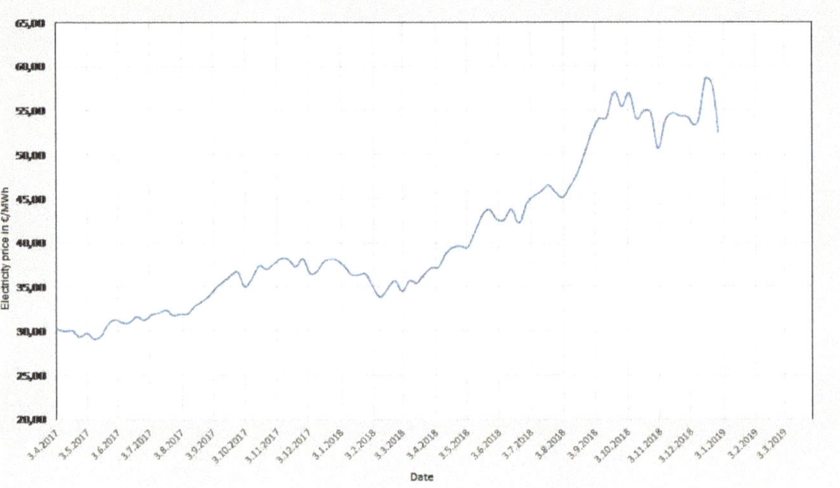

Figure 3. Electricity end user prices development on the stock market—the index PXE (Source: [33]).

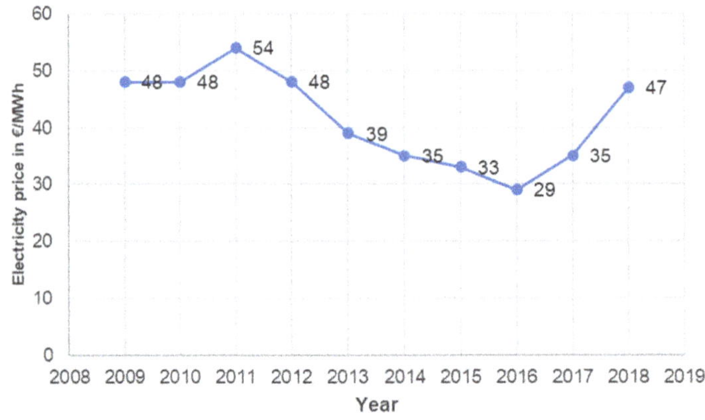

Figure 4. Stock market year-on-year increase of average electricity end user prices (€/year) in period 2009–2018 (Source: [33]).

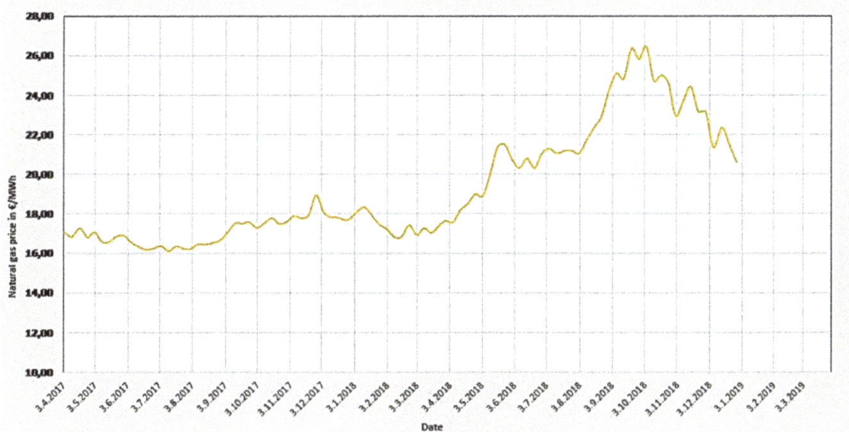

Figure 5. Gas end user prices on the stock market—the index PXE (Source: [33]).

Figure 3 shows the development of the electricity price on the stock market—the PXE index in the period April 2017–January 2019.

The stock market average electricity end user prices for period 2009–2018 are introduced in Figure 4.

Recently, the price of electricity had been significantly rising, being mainly driven by the course of electricity prices in neighbouring countries, as well as gas price developments and the price of greenhouse gas emission allowances (Figures 5 and 6).

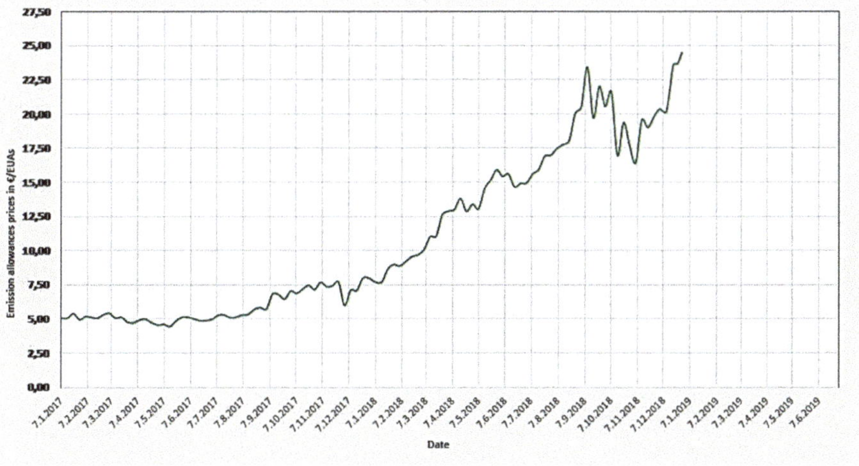

Figure 6. End user prices of emission allowances for greenhouse gases (Source: [34]).

Figure 7 shows the course of the spot price of electricity in Slovakia in 2018 according to the Energetika.cz web portal [19].

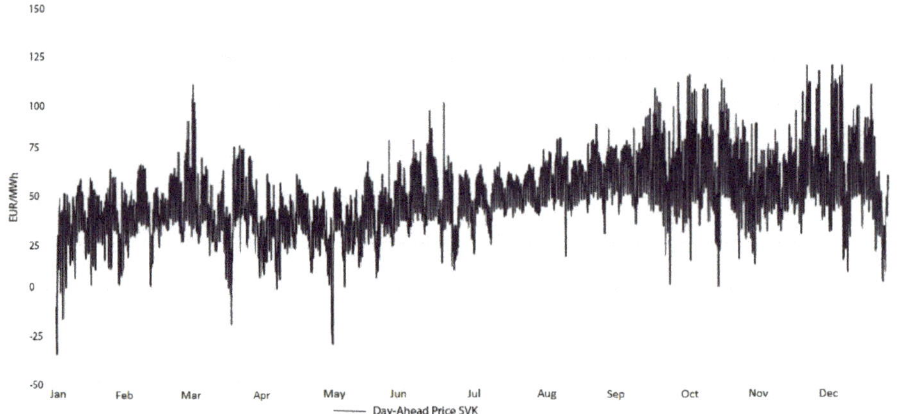

Figure 7. Electricity spot prices in 2018 (Source: [33]).

The rising price trend in 2018 was due to the extraordinary rise in commodity prices on stock markets. The prices that are introduced in Figure 7 reflect the year-on-year increase of prices in the production of electricity by 27%, gas by 24%, and emission allowances increase by more than 200%.

Figure 8 introduces the utilization of the Slovak electrical system and its coverage by the energy sources on 7.12.2018.

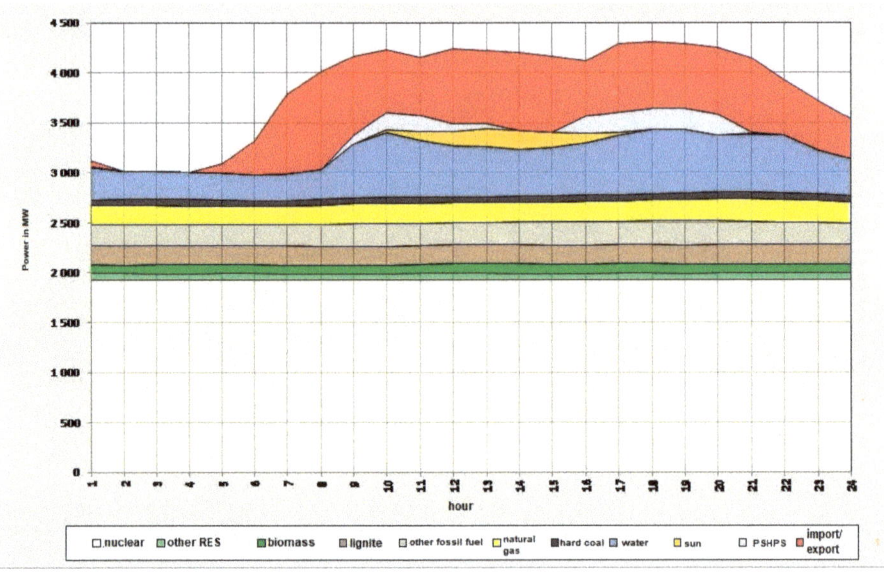

Figure 8. Slovak electrical system utilization and its coverage by the energy sources on 7.12.2018 (Source: [35]). PSHPS—Pumped-storage hydropower stations.

It can be seen from Figure 7 that, during the selected day, the least dynamic variation in electricity (power) produced was recorded by the nuclear power plants, sources of electricity from biomass and other RES, lignite and hard coal, other fossil fuels, and natural gas. On the other hand, significantly dynamically variable electricity production (power) is evident in the case of hydropower plants, pumped-storage hydropower stations (PSHPS), solar power stations, and import/export. It is also clear from the figure that, during daylight, when the power of solar power plants increases, hydropower plants and PSHPS simultaneously control this power.

5. Heat Supply from Power Plants of Slovenske Elektrarne, Inc.

Slovenské elektrarne (SE, Inc.)/Slovak Power Plants is the third largest heat supplier in Slovakia. In 2017, SE produced 898 GWh and sold 705 GWh of heat. The largest system of central heat supply (CHS) within the SE is the system in the Jaslovske Bohunice nuclear power plant. The heat that is produced by this system supplies the towns Trnava, Hlohovec, Leopoldov and Jaslovske Bohunice. In addition to the supply of non-residential premises, institutions, public buildings, and medical and school facilities, the heat is supplied for heating and water to 22,730 dwellings, including 364 family houses in the village Jaslovske Bohunice. The second largest CHS system belongs to the Novaky heating power plant and it supplies about 11,494 flats in the towns of Prievidza, Novaky, and Zemianske Kostolany with heat for heating and water heating, including 46 family houses in Zemianske Kostolany.

The highest electricity production used for heating during the year occurs during the heating season (November—March), see Figure 9.

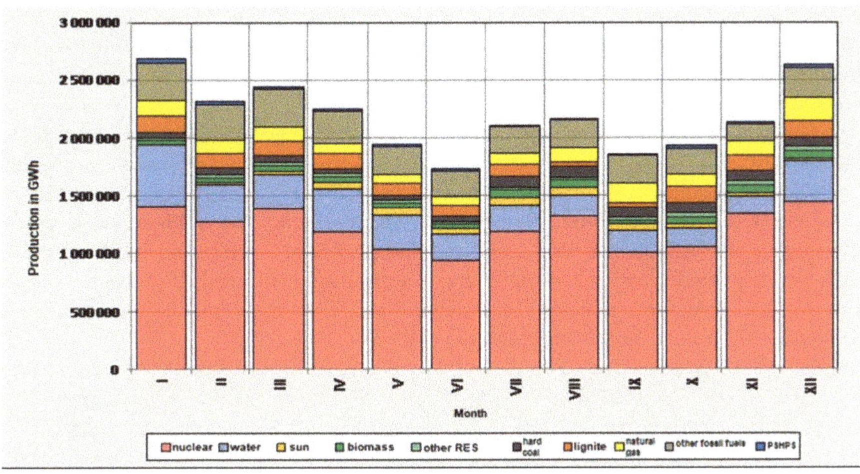

Figure 9. Electricity production for heating at sources in Slovakia in 2018 (Source: [35]).

6. Environmental Consequences of Electricity Production

Coal mining has traditionally been an important industry in the Slovak Republic. Hard coal is mostly used in the steel industry. Lignite is mostly used for power and heat generation at the Novaky power plant. In 2016, the power plant accounted for 6.4% of the total power generation in the country. The remaining extractable lignite reserves in the Slovak Republic would be sufficient to supply the Novaky power plant for at least for another 20 years. [36]

However, in recent decades, coal has moved from the largest energy source in the country's energy production, total primary energy supply (TPES), and electricity generation to only account for around 10–20% of the energy mixes. Starting the operation of nuclear power and the growth in bioenergy use has reduced the relevance of coal for the Slovak Republic's energy security, and some uneconomical coal mines are now closed down. The remaining coal mines depend on subsidies for domestic coal use in power generation. [36]

According to [36], two-thirds of coal supply is imported, mainly from the neighbouring countries and the Russian Federation. Russia accounted for 29% of total imports, followed by the Czech Republic (25%) and Poland (20%). Most of the imported coal is hard coal, but lignite also accounted for 11% of the total coal imports in 2017. Coal imports declined by 31% from 2007 to 2017, which is in line with the decline in coal demand. Domestic lignite production accounts for 30% of the total coal supply when measured in weight, and 14% when measured in energy content. Lignite production decreased by 13% in the past decade and several coal mines are now closed.

The challenge is that domestic lignite is uncompetitive in power generation. However, its perceived security of supply benefits means that the government has adopted a compensation system for its use in electricity generation. Electricity from domestic lignite also benefits from a priority dispatch to the network, just as renewable energy. The subsidy costs electricity users around 100 million EUR a year, or around 14,000 EUR per employee in coal mining and related services. [37]

The government considers that an abrupt termination of coal mining in the Upper Nitra would radically increase unemployment in the area. It prefers to find solutions that gradually phase out coal mining and the subsidy to minimise the impact on employment. Subsidies are also granted to mine closures. From 2002 to 2015, a total subsidy of 6.7 million EUR was given to the mining company Bana Dolina a.s., to cover the exceptional costs to end the mining activities and to cover the severance payments to workers. The closure of the Cigel mine is expected to cost around six million EUR in state aid. [37]

According to [37], practically all domestic coal is used for power generation, and therefore it is part of the electricity security equation and its future should be considered in this context. The International Energy Agency notes that, to the extent that domestic lignite is currently needed for the security of supply, this concern will be alleviated as soon as the first 470 MW nuclear unit comes on line, which is planned for completion in 2018.

There is a long-standing unacceptable situation from point of view of the environmental protection and health and safety of the population in Slovakia, because the territory of Slovakia is mostly affected by the emissions arising from the cheapest electricity production in neighbouring countries, not as much from the Slovak electricity production sources and coal mining. The problem of coal-fired power production is not the technologies used to produce the electricity that must meet the strict emission limits that are set by the EU and Slovak legislation. The problem is the method of coal mining and the preparation of fuel, in general. In neighbouring countries, the environmentally unacceptable surface mining of coal is the main source of environment pollution.

It is evident that "responsible persons", not only activists, are aware of the effects and consequences of dust scatter, so the fuel dumps have to be covered by the Best Available Techniques reference documents (BREF) for the best available techniques (BAT) and the ash storage under a constant water level, but hundreds of square kilometres of surface mines remain without the corresponding technical solution.

Surface mines in Poland, Germany, and the Czech Republic pose a threat to the environment, especially since hundreds of square kilometres of deposit areas are exposed to moving air, which directly carries free dust particles into the breathable layer. The micro particles of dust from the area of thousands of square kilometres of surface mines in Germany, Poland, and the Czech Republic are issued by the movement of air directly into the layers of breathable air to air pollutants, such as particulate pollutants causing smog situation, not only in their countries of origin, but also in Slovakia.

Coal mining needs to be disrupted and relocated, because it is 10 times more overburdened than the coal itself. When mining 34 million tons of coal per year, 340 million tons of overburden is moved, while using the dunes and open conveyor belts. When processing and transporting, the moving air generates very dangerous solid pollutants that are emitted from the overburden layer in millions of tons. This dust type can reach the share even more than 1%.

Appendix A (Table A1) presents a list of selected sources of electricity production of the Slovak power system. The installed power, production, own electricity consumption, and emissions in 2017 are updated by expert estimation for 2018. For non-emission electricity producers, a negative emission value is shown in the table, which represents the saved CO_2 emissions. The biggest savings are in the nuclear power plants (about 14 million tons of CO_2) and hydropower plants (4.2 million tons of CO_2). Electricity production in Slovakia is associated with the annual CO_2 emission of 8.2 million tons of CO_2, representing 324 kg/MWh. The utilization of the electricity in Germany and in the Czech Republic is estimated to be more than 700 kg/MWh in 2018. Electricity utilization in Poland is estimated to be about 1000 kg/MWh. With such utilizations, the development of electro mobility in power production would cause an increase in greenhouse gas production.

7. Conclusions

Energy security is being tackled under the influence of energy in social life and in the economy, particularly in the liberalization of energy markets, in diversification and in resource efficiency, and in procedures that multiply competition in energy markets. More broadly, there are national and international measures to ensure stable energy supplies at affordable and stabilized prices, but also to guarantee the protection of critical energy infrastructure and to demonstrate state readiness to effectively respond to potential crisis. Since energy is one of the factors that determine the inputs into production and thus production costs, rising energy prices are multiplying pressures on producers, reducing profit. Therefore, it is necessary to secure energy resources in a liberalized market at acceptable prices in the interests of economic development.

From this point of view, the aim of this article was to summarize the information on EU and Slovak energy and environmental legislation, describe the current energy sources that are used for electricity production, current state of electricity production, consumption, and end user prices development. This information was also supplemented with information on heat production from energy sources and environmental consequences of lignite completed with information on the environmental consequences of electricity production using lignite (surface mining) as an energy source.

The data and charts used in the article come from publicly available sources.

Author Contributions: Conceptualization, J.J.; methodology, A.M., M.L. and M.T.; formal analysis, J.J., M.T.; resources, J.J., M.T.; data curation, A.M., M.L.; writing—original draft preparation, A.M.; writing—review and editing, A.M.; project administration, M.L..; funding acquisition, M.L.

Funding: This research was funded by the KEGA Grant Agency under the project KEGA 013TU Z-4/2017.

Conflicts of Interest: The authors declare no conflict of interest. The funders had no role in the design of the study; in the collection, analyses, or interpretation of data; in the writing of the manuscript, or in the decision to publish the results.

Appendix A

Table A1. List of selected sources of electricity production within the Slovak electrical system. (Source: [31,32]).

Source	Installed Power (MW)	Electricity Supply (MWh)	Own Consumption (%)	Emissions CO_2 (t)	Emissions NO_x (t)	Emissions SO_x (t)
Diesel generator station Levice	32.000	0		0	0	0
Diesel generator station Moldava	32.000	0		0	0	0
Diesel generator station Sucany	32.000	0		0	0	0
Diesel generator station PPC Trakovice	10.000	93	19.33	104	0	0
Diesel generator stations	**106.000**	**93**	**19.33**	**104**	**0**	**0**
Nuclear power plant Bohunice	1000.000	7,059,256	6.88	−7,294,478	x	x
Nuclear power plant Mochovce	940.000	7,000,331	6.67	−6,752,350	x	x
Nuclear power plants	**1940.000**	**14,059,588**	**6.77**	**−14,046,828**	**x**	**x**
Hydroelectric power plant Orava	21.750	26,893	1.02	−24,454	x	x
Hydroelectric power plant Tvrdosin	6.100	11,205	4.37	−10,545	x	x
Hydroelectric power plant Cierny Vah	735.160	234,765	1.32	−214,125	x	x
Hydroel. power plant Liptovska Mara	198.000	103,967	1.89	−95,377	x	x
Hydroelectric power plant Krpelany	24.750	55,475	1.94	−50,913	x	x
Hydroelectric power plant Sucany	38.400	78,117	1.06	−71,060	x	x
Hydroelectric power plant Lipovec	38.400	70,422	1.35	−64,245	x	x
Hydroelectric power plant Zilina	72.000	142,641	0.63	−129,195	x	x
Hydroelectric power plant Hricov	31.500	53,442	0.31	−48,246	x	x
Hydroelectric power plant Miksova	93.600	154,285	0.53	−139,598	x	x
Hydroelectric power plant P. Bystrica	55.200	93,793	0.91	−85,185	x	x
Hydroelectric power plant Nosice	67.500	143,161	0.26	−129,180	x	x

Table A1. *Cont.*

Source	Installed Power (MW)	Electricity Supply (MWh)	Own Consumption (%)	Emissions CO$_2$ (t)	Emissions NO$_x$ (t)	Emissions SO$_x$ (t)
Hydroelectric power plant Ladce	18.900	78,650	1.07	−71,549	x	x
Hydroelectric power plant Ilava	15.000	73,292	1.26	−66,808	x	x
Hydroel. power plant Dubnica/Vahom	16.500	72,605	1.11	−66,076	x	x
Hydroelectric power plant Trencin	16.100	69,433	0.29	−62,669	x	x
Hydroelectric power plant Kostolna	25.500	96,867	1.26	−88,297	x	x
Hydroel. power plant N. M./Vahom	25.500	98,271	0.47	−88,862	x	x
Hydroel. power plant Horna Streda	25.500	97,937	1.42	−89,414	x	x
Hydroelectric power plant Madunice	43.200	127,370	0.95	−115,730	x	x
Hydroelectric power plant Kralova	45.060	96,182	0.21	−86,747	x	x
Hydroelectric power plant Ruzin I.	60.000	55,038	2.36	−50,732	x	x
Hydroel. power plant V. Kozmalovce	5.320	9834	4.17	−9236	x	x
Hydroelectric power plant Domasa	12.400	9211	1.66	−8429	x	x
Hydroelectric power plant Dobsina I.	24.000	27,688	1.96	−25,418	x	x
Hydroelectric power plant Gabcikovo	720.000	1,430,006	1.02	−1,300,331	x	x
Hydroelectric power plant Cunovo	24.280	109,706	2.33	−101,094	x	x
Small hydroel. power plants < 5.0 MW	79.380	968,044	3.92	−906,786	x	x
Hydroelectric power plants	**2539.000**	**4,588,301**	**1.69**	**−4,200,300**	x	x
Solar power plants	**530.000**	**609,000**	**0.00**	**−548,100**		
Waste incinerator OLO Bratislava	6.300	35,350	23.71	−125,102	163	642
Waste incinerator Kosit Kosice	6.000	33,096	25.79	−120,409	157	618
Municipal waste	**12.300**	**68,446**	**24.73**	**−245,511**	**321**	**1260**
CHP source, Bioenergy, s.r.o. Bardejov	8.200	69,067	0.00	−111,889	162	0
CHP source, Bioenergy, Topolcany	8.200	69,410	0.00	−99,488	148	0
CHP source, Bukocel a.s. Hencovce	18.000	59,915	14.19	−228.513	341	0
CHP source, Energy Edge ZC, Zarnovica	11.000	97,257	15.00	−106.792	273	0
CHP source, Martinska teplarenska	9.000	12,227	15.16	−45,779	68	0
CHP source, Mondi SCP	137.600	500,979	15.00	−2,340,213	3058	0
CHP source, Zvolenska teplarenska	5.000	36,515	3.92	−90,642	188	0
CHP source, Chemes Humenne	24.000	50	15.99	−180	0	0
CHP source, DALKIA Ziar nad Hronom	15.000	21,138	34.91	−85,963	144	0
Another biomass	**236.000**	**866,559**	**13.09**	**−3,109,459**	**4384**	**0**
Biofuel	**105.000**	**603,500**	**6.00**	**−1,059,300**	**324**	**0**
CHP source, Bukocel a.s. Hencovce	1.600	1394	14.19	2753	8	3
CHP source, Martinska teplarenska	33.000	37,284	15.16	21,739	208	342
CHP source, Zvolenska teplarenska	35.000	39,373	3.92	91,613	239	664
CHP source, Zilinska teplarenska	49.768	96,485	9.18	243,765	683	268
Brown-coal fired Elektrarne Novaky	266.000	1,479,820	10.85	1,656,771	1546	2568
Lignite	**385.368**	**1,654,356**	**10.71**	**2,016,641**	**2684**	**3845**
CHP source, CMEPS, Slovnaft	89.000	288,922	21.83	764,693	1124	442
CHP source, Chemes Humenne	24.000	743	15.99	2357	4	2
Heavy fuel oil	**113.000**	**289,664**	**21.81**	**767,051**	**1125**	**443**
CHP source, Chemes Humenne	24.000	11,858	15.99	45,204	71	269
CHP source, Teplaren Kosice, a.s.	65.000	152,510	8.33	416,555	656	2482
CHP source, Teplaren USS Kosice	67.000	340,956	16.00	1,034,027	1627	6160
CHP source, DALKIA Ziar nad Hronom	10.000	6605	34.91	32,496	51	194
Brown-coal-fired Elektrarne Vojany SE	220.000	376,998	11.65	537,655	155	3562
Hard coal	**386.000**	**888,927**	**13.13**	**2,065,937**	**2560**	**12,667**
CHP source, Chemes Humenne	24.000	3477	15.99	5362	14	0
CHP source, Teplaren Kosice, a.s.	56.000	94,832	8.33	192,979	408	9
CHP source, Teplaren USS Kosice	45.000	23,517	16.00	54,403	112	2
CHP source, Mondi SCP, a.s.	3.336	52	15.00	119	0	0
CHP source, DALKIA Ziar nad Hronom	4.400	5294	34.91	10,537	28	1
CHP source, BAT - Teplaren Vychod	24.500	77,525	9.12	259,321	516	11
CHP source, BAT - Teplaren Zapad	25.000	39,358	8.08	125,971	251	5
CHP source, GT Teplaren Radvan	5.200	16,198	0.89	30,490	64	1
CHP source, GT1	6.300	46,944	1.18	92,311	190	4
CHP source, KGJ TEDOM	2.606	10,045	8.74	21,390	44	1
CHP source, PPC Investment Bratislava	217.960	1354	1.90	2683	6	0
CHP source, PPC Energy Bratislava GT	58.000	260	5.27	533	1	0
CHP source, PPC Levice	87.000	329,590	4.07	649,854	1362	29
CHP source, PPC Povazska Bystrica	57.966	258,131	6.00	249,213	494	11
CHP source, Energochem SVIT	11.970	58,461	3.22	63,804	127	3
CHP source, Zilinska teplarenska	49.768	172	9.18	392	1	0
CHP source, CGU, totally 108 sources	85.000	350,800	4.41	326,336	658	14
Dante Power Plant	50.000	1581	0.00	901	4	0
Natural gas	**814.006**	**1,317,593**	**5.60**	**2,086,599**	**4281**	**92**
CHP source, Teplaren USS Kosice	85.000	426,355	16.00	1,293,017	2035	7703
Metallurgic gases	**85.000**	**426,355**	**16.00**	**1,293,017**	**2035**	**7703**
Energy sources of the ES SR	**7245.674**	**25,339,286**	**6.81**	**8,229,349**	**12,688**	**24,751**

References

1. An Energy Policy for Europe. Available online: http://www.europarl.europa.eu/meetdocs/2004_2009/documents/com/com_com(2007)0001_/com_com(2007)0001_en.pdf (accessed on 14 January 2019).
2. Energy Efficiency Action Plan 2007–2010. 2007. Available online: https://eur-lex.europa.eu/LexUriServ/LexUriServ.do?uri=COM:2011:0109:FIN:EN:PDF (accessed on 13 January 2019).
3. Strategic Energy Technology Plan. 2007. Available online: https://ec.europa.eu/energy/en/topics/technology-and-innovation/strategic-energy-technology-plan (accessed on 23 January 2019).
4. Third Liberalization Package. 2007. Available online: http://www.europarl.europa.eu/ftu/pdf/en/FTU_2.1.9.pdf (accessed on 23 January 2019).
5. Climate-Energy Package. 2008. Available online: https://ec.europa.eu/clima/policies/strategies/2020_en (accessed on 6 April 2019).
6. Energy Efficiency Plan 2011. Available online: https://ec.europa.eu/clima/sites/clima/files/strategies/2050/docs/efficiency_plan_en.pdf (accessed on 11 January 2019).
7. Second Strategic Energy Review—An EU Energy Security and Solidarity Action Plan. Available online: https://eur-lex.europa.eu/legal-content/EN/TXT/PDF/?uri=CELEX:52008DC0781&from=DE (accessed on 11 January 2019).
8. European Economic Recovery Plan. 2008. Available online: https://ec.europa.eu/inea/sites/inea/files/download/publications/eerp_v11.pdf (accessed on 13 January 2019).
9. Treaty of Lisbon. Available online: https://eur-lex.europa.eu/legal-content/EN/TXT/HTML/?uri=CELEX:12007L/TXT&from=EN (accessed on 11 January 2019).
10. Europe 2020 Strategy. Available online: http://ec.europa.eu/eu2020/pdf/COMPLET%20EN%20BARROSO%20%20007%20-%20Europe%202020%20-%20EN%20version.pdf (accessed on 13 January 2019).
11. Energy 2020 A Strategy for a Competitive, Sustainable and Secure Energy. Available online: https://eur-lex.europa.eu/legal-content/EN/TXT/HTML/?uri=LEGISSUM:en0024&from=EN (accessed on 12 January 2019).
12. Energy Efficiency Directive 2012/27/EU. Available online: https://eur-lex.europa.eu/legal-content/EN/TXT/PDF/?uri=CELEX:32012L0027&rid=3 (accessed on 11 January 2019).
13. Communication "Energy Infrastructure. Priorities for 2020 and Beyond". Available online: https://ec.europa.eu/energy/sites/ener/files/documents/2011_energy_infrastructure_en.pdf (accessed on 12 January 2019).
14. Regulation EU No. 347/2013 on Guidelines for Trans-European Energy Infrastructure. Available online: https://eur-lex.europa.eu/legal-content/EN/TXT/HTML/?uri=CELEX:32013R0347&from=en (accessed on 11 January 2019).
15. Regulation EU No. 1316/2013 Establishing the Connecting Europe Facility. Available online: https://eur-lex.europa.eu/legal-content/EN/TXT/HTML/?uri=CELEX:32013R1316&from=EN (accessed on 11 January 2019).
16. Roadmap to a Competitive Low Carbon Economy in 2050. Available online: https://ec.europa.eu/clima/sites/clima/files/strategies/2050/docs/roadmap_fact_sheet_en.pdf (accessed on 11 January 2019).
17. Emisie.sk. Available online: http://emisie.icz.sk/ (accessed on 19 January 2019).
18. A 2030 Framework for Climate and Energy Policies. Available online: https://eur-lex.europa.eu/legal-content/EN/TXT/PDF/?uri=CELEX:52013DC0169&rid=8 (accessed on 13 January 2019).
19. 2030 Climate and Energy Framework. Available online: https://www.iddri.org/sites/default/files/import/publications/cs-iddri-eu-2030-framework_energy-security.pdf (accessed on 12 January 2019).
20. Communication "Progress towards Completing the Internal Energy Market". Available online: http://data.consilium.europa.eu/doc/document/ST-16037-2014-INIT/en/pdf (accessed on 12 January 2019).
21. Clean Energy for All Europeans. Available online: https://ec.europa.eu/energy/en/topics/energy-strategy-and-energy-union/clean-energy-all-europeans (accessed on 6 April 2019).
22. 2050 Long-Term Strategy. Available online: https://ec.europa.eu/energy/en/topics/energy-strategy-and-energy-union/2050-long-term-strategy (accessed on 6 April 2019).
23. Act No. 251/2012 Coll. on Energy and on Amendments to Certain Acts. Available online: http://www.zakonypreludi.sk/zz/2012-251 (accessed on 7 February 2019).
24. Act No. 250/2012 Coll. Regulation on Network Industries. Available online: http://www.zakonypreludi.sk/zz/2012-250 (accessed on 7 February 2019).
25. Act No. 476/2008 Coll. on Energy Efficiency. Available online: https://www.zakonypreludi.sk/zz/2008-476 (accessed on 6 April 2019).

26. Act No. 309/2009 Coll. on the Promotion of Renewable Energy Sources and High Efficiency Cogeneration and on Amendments to Certain Acts. Available online: http://www.zakonypreludi.sk/zz/2009-309 (accessed on 7 February 2019).
27. Decree of the Office for Regulation of Network Industries no. 24/2013 Coll., Laying down the Rules for the Functioning of the Internal Market in Electricity and the Rules for the Functioning of the Internal Gas Market. Available online: http://www.zakonypreludi.sk/zz/2013-24 (accessed on 7 February 2019).
28. Decree of the Ministry of Economy No. 599/2009 Coll. Aimed at Implementing the Provisions Mentioned in the Act on the Promotion of Renewable Energy Sources and High Efficiency Cogeneration. Available online: https://www.zakonypreludi.sk/zz/2009-599 (accessed on 6 April 2019).
29. Decree of the Office for Regulation of Network Industries no. 80/2015 Coll. Amending the Decree of the Regulatory Office for Network Industries No. 490/2009 Coll. Laying down Details on the Promotion of Renewable Energy Sources, High Efficiency Cogeneration and Biomethane. Available online: https://www.zakonypreludi.sk/zz/2015-80 (accessed on 6 April 2019).
30. Decree of the Office for Regulation of Network Industries no. 189/2014 Coll. Amending the Decree of the Regulatory Office for Network Industries No. 221/2013 Coll. Laying down the Price Regulation in the Electricity Industry. Available online: https://www.zakonypreludi.sk/zz/2014-189 (accessed on 6 April 2019).
31. Yearbook 2017 of the SED. Available online: https://www.sepsas.sk/Dokumenty/RocenkySed/ROCENKA_SED_2017.pdf (accessed on 21 January 2019).
32. URSO. Available online: http://www.urso.gov.sk/?language=en (accessed on 21 January 2019).
33. KBP Praha. Available online: http://www.kbp.cz/ (accessed on 23 January 2019).
34. KBB Bratislava. Available online: http://www.sk.kbb.sk/ (accessed on 23 January 2019).
35. Oenergetice.cz. Available online: https://oenergetice.cz/ (accessed on 20 January 2019).
36. Energy Policies of IEA Countries. Available online: https://www.connaissancedesenergies.org/sites/default/files/pdf-actualites/energy_policies_of_iea_countries_slovak_republic_2018_review.pdf (accessed on 29 April 2019).
37. Energy Policy of the Slovak Republic. Available online: http://www.economy.gov.sk/uploads/files/47NgRIPQ.pdf (accessed on 28 April 2019).

Review

Smart Energy Management Policy in India—A Review

Komali Yenneti [1], Riya Rahiman [2], Adishree Panda [2] and Gloria Pignatta [1,*

[1] Faculty of Built Environment, University of New South Wales (UNSW), Sydney 2052, Australia
[2] Centre for Urban Planning & Governance, TERI, New Delhi 110003, India
* Correspondence: g.pignatta@unsw.edu.au

Received: 29 June 2019; Accepted: 17 August 2019; Published: 21 August 2019

Abstract: India accounts for six per cent of the world's primary energy consumption. Rapid urbanization and rapid urban population growth have had a serious impact on energy consumption and subsequent carbon emissions. In particular, cities face a complex and interrelated set of challenges across different sectors (building environment, mobility, water and waste management and public services). Re-examining these challenges by integrating smart energy management (SEM) principles is critical for sustainable and low-carbon urban development. In addition, managing energy footprint is one of the most challenging goals for cities, and as existing cities evolve and transform into smart cities, SEM becomes an integral part of the urban transformation. This article comprehensively reviews the different SEM technologies for different sectors (construction, transportation, public services, water and waste), the policies, and the current challenges and opportunities for SEM policy governance in India. Making urban energy smart can manage a city's energy footprint and have a positive impact on future carbon emissions.

Keywords: smart cities; smart energy management; India; energy efficiency; low-carbon mobility; water-energy nexus

---

1. Introduction

1.1. Background

India is rapidly urbanizing. Urban development is one of the main drivers of energy demand and consumption. The nation accounts for 18 per cent of the world's population and 6 per cent of global primary energy consumption [1]. India's urban population is expected to grow from 31.6 per cent to 57.7 per cent by 2050, with additional impacts on energy consumption and carbon emissions [2]. The task of managing and reducing energy demand, energy consumption and energy-related carbon emissions is often a challenge for urban planners [3,4].

In this context, it is important to address urban challenges such as sustainable transportation, access to reliable and clean energy, green and resilient infrastructure, and waste management through the principles of smart energy management (SEM). In particular, there is a need to integrate SEM principles in different sectors, such as buildings, transport, water, waste and public services, to reduce carbon emissions and achieve sustainable development goals (SDGs). This integrated SEM will also ensure the optimal utilization of available resources and reduce reliance on unsustainable energy sources.

To this end, The Energy and Resources Institute (TERI) and the University of New South Wales (UNSW) organized Australia-India knowledge exchange workshops on "Smart Energy Management for Sustainable Cities" in Sydney and Delhi. Sectoral experts and urban practitioners from local urban institutions, policy think tanks, academia and research institutions, as well as industry participated in the workshops. Through panel discussions chaired by sector experts, participants considered the way forward for integrated SEM in Indian cities. This paper combines the main results of the workshops and a comprehensive literature review to introduce the state of policy on SEM in India. It highlights SEM governance, SEM strategies and initiatives in different sectors, and potential challenges, drivers and opportunities for SEM in India. The review will contribute to knowledge creation and understanding on the SEM policy landscape in India.

1.2. The Definition of SEM

In order to build an understanding of SEM, it is important to first become familiar with the concepts and principles of energy management. Energy management can be defined as the science of planning, guiding, and controlling energy supply and consumption to maximize productivity and comfort, and minimize energy costs and pollution [5]. Simply put, energy management involves the conscious, wise and efficient use of energy. Energy management also requires the process of monitoring, controlling and saving energy, optimized energy utilization, management of energy resources, and active energy efficiency. With the rapid expansion of India's urban areas, managing the energy footprint is becoming a challenging target for cities. Therefore, energy management has recently become an integral part of urban transformation.

Taking into account local resources and the needs of stakeholders, smart cities are expected to become more autonomous and manage their energy footprints more effectively. In this paper, SEM entails energy management as an important component of smart cities. A smart city is a sustainable and efficient urban centre designed to provide a high quality of life for its residents by optimizing resources [6]. In this article, SEM is defined as follows:

> "Smart energy management is a component of smart city development aiming at a site-specific continuous transition towards sustainability, self-sufficiency, and resilience of energy systems, while ensuring accessibility, affordability, and adequacy of energy services, through optimized integration of energy conservation, energy efficiency, and local renewable energy sources. It is characterized by a combination of technologies with information and communication technologies that enables integration of multiple domains and enforces collaboration of multiple stakeholders, while ensuring sustainability of its measures." [7] (p. 57)

The mainstreaming of SEM involves the integration of energy technology systems, enabling policies, strategies, institutional change, awareness, training and capacity-building programmes, energy conservation measures and energy audits.

2. SEM Policy Governance in India

The Government of India (GoI) has been developing policies and programmes to guide the mainstreaming of SEM into urban planning. To meet citizens' energy needs and reduce carbon emissions, the Indian government has adopted a two-pronged approach, focusing on supply and demand. In the area of power generation, the focus is on greater use of renewable energy (solar and wind). In the current policy landscape, the impetus to reimagine national energy supply and demand management through technological advances is in place. In addition, in accordance with the intended nationally determined contributions (INDCs), a strategy has been developed to reduce the emission intensity of the GDP by 33–35 per cent by 2030 from 2005 levels and to achieve 40 per cent fossil-free power generation capacity by 2030 [8]. On the demand side, efforts are being made to improve energy efficiency through innovative policy measures that fall within the scope of the Energy Conservation Act of 2001. In this section, some of the key SEM policies are discussed.

2.1. The Energy Conservation Act (EC Act), 2001, and Amended in 2010

The Energy Conservation Act 2001 (EC Act), as amended in 2010, was enacted by the Ministry of Electricity (MoP) to reduce energy intensity. In 2002, the Bureau of Energy Efficiency (BEE) was established as the central statutory body for the implementation of the EC Act. The Act regulates the Energy Conservation Building Code (ECBC), standards and labelling of equipment and electrical appliances, and energy consumption norms for energy-intensive industries. It is estimated that energy consumption through the implementation of the ECBC will be reduced by at least 50 per cent by 2030.

In addition, the Act requires the GoI and BEE to promote energy efficiency in all sectors. The law also requires designated agencies at the state level to enforce the law and improve energy efficiency. Through the Act, the BEE has launched a series of energy-saving initiatives in the areas of home lighting, commercial buildings, standards and home appliance labels. For example, the Energy Efficiency Labeling Program is designed to reduce equipment energy consumption without compromising on performance. The law also sets minimum energy standards for new commercial buildings across India.

2.2. National Mission on Enhanced Energy Efficiency, 2010

The National Mission on Energy Efficiency (NMEE), launched under the National Climate Change Action Plan on Climate Change (NAPCC), was launched under the EA 2001. The Energy Efficiency Services Limited (EESL), established as a business entity under the mission, provides market leadership. The goal of the NMEEE is to meet the country's energy needs and provide energy efficiency. During the full implementation phase, the NMEEE intends to save approximately 23 million tons of fuel per year and reduce carbon emissions by 95.55 million tons per year [9]. The first cycle of the Perform, Achieve and Trade (PAT) initiative (2012–2015) reduced carbon emissions by 31 million tons, or about 1.93 per cent of India's total emissions [9]. While the BEE recognizes the NMEE as one of its key tasks, PAT entails effective investment mechanisms and financial instruments, such as energy efficiency certificates. In addition, the mission pays greater attention to inter-sectoral linkages and close coordination between energy demand and supply sectors.

2.3. National Mission for Sustainable Habitat, 2010

The National Mission on Sustainable Habitat (NMSH), envisaged in the NPACC, aims to integrate climate change adaptation and mitigation into urban planning. This is achieved by improving the energy efficiency of buildings, managing solid waste and switching to public transport. The code estimates that energy consumption will be reduced by 50 per cent by 2030. In line with this mission, BEE has also developed energy efficiency design guidelines for multi-storey residential buildings in composite and hot and dry climates. In the area of green buildings and energy efficiency, a green building code has been incorporated into the workbook of the Central Works Department (CPWD). Draft building by-laws were also published to help state and territory governments incorporate their necessary provisions into their respective building by-laws. A number of activities envisaged under the NMSH were included in the Atal Recovery and Urban Transformation Mission (AMRUT). Examples include the implementation of pedestrian, non-motorized and public transport facilities, as well as the establishment and upgrading of urban green space, parks and recreation centres.

2.4. National Solar Mission, 2010

The National Solar Mission (NSM), launched under the NAPCC, envisages a comprehensive strategy for phased development of grid-connected and off-grid solar applications. According to the Ministry of New and Renewable Energy (MNRE), the aim is to provide an enabling environment for the penetration of solar technology in the country. The initial goal of the NSM was to generate 20 gigawatts (GW) grid-connected solar power. In 2014, the target was upscaled from 20 GW to 100 GW by 2022. It has separate goals and targets for solar cities, solar water heating, solar street lighting, solar cooking, refrigeration and other off-grid applications and rooftop systems. The Solar Cities programme is designed to incentivize each city to reduce projected demand for traditional energy by at least 10 per cent within five years.

2.5. Smart Cities Mission, 2015

The Ministry of Housing and Urban Affairs (MoHUA) launched the Smart Cities Mission in 2015 to address governance and infrastructure concerns of the growing urban population of cities and to promote quality of life. By integrating the internet of things (IoT) technology, the mission aims to integrate smart technologies in different sectors to achieve sustainable urban development in an energy-efficient and cost-effective manner. The smart city proposals entailed two major aspects: (1) a pan city project, which includes information and communication technology (ICT) solutions such as an integrated traffic management system (ITMS), automated city level waste collection, e-governance and rooftop solar panels in government buildings; and (2) an area-based development (ABD) project, wherein a small area within the city is identified and is developed with identified smart solutions for improving its infrastructure, which then becomes a prototype for other parts of the city. The mission aims to create energy efficient urban spaces to reduce the burden on existing resources. As per the mission guidelines, 80 per cent of the buildings in the smart cities need to be energy efficient with a 'green building' design and 10 per cent of a smart city's energy demand should be met by solar energy.

2.6. Draft National Energy Policy, 2017

The National Energy Policy (NEP) prepared by NITI Aayog focuses on energy access for all, reducing dependence on fossil-fuel imports, promoting low-carbon development and sustainable economic growth. Considering energy is handled by several ministries, this policy aims to achieve national energy objectives through coordination between different ministries. The NEP aims to mainstream energy technologies and provide consumers with energy choices and prepare the Indian economy to be 'energy ready' by 2040.

Demand-based strategies relevant to integrating energy management for sustainable cities have been proposed under the NEP for transport, buildings and household (cooking) sectors. Under the transport sector, strategies such as transit oriented development, modal shift to rail-based mass transport systems, promotion of electric and hybrid vehicles, and ICT solutions have been proposed. Under the buildings sector, a shift towards energy efficient building materials, enforcement of ECBC, better urban planning, adoption of high efficiency lighting technologies and appliances, and under the housing sector, a shift towards modern fuels for cooking and improvement in the efficiency of cooking fuels and stoves has been proposed.

2.7. Draft National Cooling Action Plan, 2018

India mostly experiences a tropical climate and with its cities undergoing rapid urbanization, cooling requirements are projected to rise by 2.2 to 3 times (2017 baseline) in the next decade [10]. The implications of this increase will be significant through increased power generation capacity, peak load and carbon footprint. It is against this background that the Ministry of Environment, Forests and Climate Change (MoEFCC) has developed a draft National Cooling Action Plan (NCAP). The NCAP aims to promote sustainable and intelligent cooling practices across the country over the next 20 years. It covers cooling growth scenarios and strategies, transportation cooling, cold chain and refrigeration, air conditioning and refrigeration technologies, service sectors, and cross-cutting policies and regulations.

3. SEM in Different Sectors

Cities are complex ecosystems that cover a wide range of sectors, including construction, transportation, water, waste and public services. To understand the potential of integrated SEM for Indian cities, the workshop in Delhi focused on identifying challenges and opportunities for the adoption of SEM. Discussions focused on existing policies and strategies applicable to different sectors. Building on the results of the workshop and further comprehensive literature review, this section identifies SEM solutions and strategies for each sector, as well as the challenges for the adoption of SEM in Indian cities.

3.1. Enhancing Sustainable Energy Management of Buildings

Old or new, public or private owned, commercial or residential, single or multi-occupant buildings are facilities where people live, work, and play. They form the landscape of a city and are home for its people. At the same time, buildings are the largest energy consumers. Buildings account for over 40 per cent of India's energy consumption—this will soon increase to nearly 60 per cent [11]. In buildings, energy services such as cooling, heating, hot water, lighting, electricity and natural gas are used daily for the safety and comfort of the occupants. These facilities account for three quarters of total Greenhouse Gas (GHG) emissions in urban areas. [6]. As a result, cities need to make energy-efficient, smarter, greener and sustainable buildings.

The main objective of SEM solutions in buildings is to minimize the environmental impact of various energy services on the building lifecycle and reduce energy costs [7]. They should be able to optimize energy consumption and demand, manage occupant comfort, and help create household energy independence that will help sustain the grid (ibid.). SEM solutions in buildings fall into three categories based on their applicability: (i) solutions that address energy consumption by providing efficient control of building energy systems; (ii) solutions that deal with energy demand response; and (iii) solutions that integrate solar passive design and sustainable materials [6].

By integrating energy generation, storage, distribution, and automation, the solutions in the first approach provides greater comfort, functionality, and flexibility. In fact, optimizing operations and managing can save 20 to 30 per cent of building energy without changing system structure or hardware configuration (ibid.). Within this approach, variable speed chillers, home temperature automation control systems and adaptive fuzzy comfort controllers are the latest focus of smart heating, ventilation, and air conditioning (HVAC) systems efforts. Lighting controls and features such as appliance control gears, day lighting integration using building information modeling (BIM) tools, occupancy sensors, fixtures with photometric characters, and light-emitting diode (LED) lamps are common smart lighting solutions [12].

Demand response is another approach. Generally, most buildings are passive consumers of energy. However, in order to achieve the expected energy objectives, the role of buildings must be transformed from passive and unresponsive energy users to active participants in energy systems [13]. This paradigm shift can be achieved through micro-grids, demand response schemes, information and control systems to manage load and consumption, and energy storage equipment [6]. In the micro-grid concept, other variants are available according to the size and type of application. Examples include nanogrids, district energy networks, combined cooling and cooling systems, and medium-scale microgrids [14].

In a passive systems approach, building insulation, thermal mass, window placement, glass type and shading are key technologies [15]. Solutions in other approaches are most effective when combined with building insulation and solar passive solutions.

Some of these smart building energy systems and strategies are already in place in Indian cities. For example, smart metering, smart grid and energy internet, rooftop solar, net metering, smart lighting, LEDs, day lighting, and smart HVAC systems are being used to achieve a smart building architecture [16]. However, due to insufficient knowledge and limited expertise [17], many other advanced solutions (e.g., smart building energy management systems, micro- and nano-grids, home automation controls) are limited to small-scale or pilot projects. The large-scale application of SEM technologies in buildings will help engineers, planners, and designers in India to achieve the lowest energy cost targets and zero environmental impact on the building life cycle [18].

For example, smart metering, smart lighting, smart grids, rooftop solar, net metering, LEDs and smart HVAC systems are being used to implement intelligent building architectures. However, due to limited knowledge and expertise, many other advanced solutions (such as smart building energy management systems, micro and nano-grids, home automation control) are limited to small or pilot projects. The large-scale application of SEM technology in buildings will help engineers, planners and designers in India achieve the minimum energy cost target with zero environmental impact on the building life cycle.

With the Smart Cities Mission and the mission of harnessing renewable energy, the future of smart energy technologies in Indian architecture is very bright, and this area will significantly contribute to the future of the technological revolution. In addition, by deploying smart technologies, conventional buildings can be transformed into smart energy buildings.

3.2. Improving the Water–Energy Nexus

In order to ensure the effective management of water, the nexus between the water and energy sectors cannot be ignored. Water is a basic requirement for meeting energy demand and supply. Evidence shows that thermal power plants account for 87.8 per cent of the country's total industrial water consumption [19]. However, the water sector currently faces several problems and challenges that hinder the effective management of water resources. For example, India accounts for 18 per cent of the world's population, but only 4 per cent of its water resources [19]. Due to limited resources, the per capita water availability is on a decline, which increases resource pressure on the country's energy requirements. There is also loss of water in urban supply systems due to inefficient distribution mechanisms. A major concern in management of the water–energy nexus is that the supply systems have been functioning independently.

An integrated approach is required to ensure that the energy and water sectors are not managed in silos. SEM of water generally refers to "a holistic approach to managing this priceless resource, and the infrastructure systems surrounding its sourcing, treatment and delivery" (Environmental Leader, 2018). SEM is needed to identify energy utilized for water consumption, supply and distribution—either for public or private usage. This will improve efficiencies in the water systems and reduce wastage.

Smart technology in the water sector usually consists of four components: (i) digital output instruments (meters and sensors), which collect and transmit information in real time; (ii) supervisory control and data acquisition (SCADA) systems, which process information and remotely operate and optimize systems; (iii) geographic information systems (GIS), which store, manage, and analyze spatial information; and (iv) software applications, which support modelling infrastructure and environmental systems by managing and reporting data to improve design, decision making, and risk management [20].

Water is a significant requirement for coal-based power plants and nuclear power plants, as well as for renewable energy production. The different stages where water is utilized indispensably include extraction and refining of fuel and in thermal production of electricity. A reservoir water supply system helps to optimize water supply levels by estimating demand [21]. Other systems that support water monitoring are the real-time hydrological data acquisition and processing systems that collect water levels, water quality, and other relevant data via satellite imaging and other communication technologies; and the generation integrated operation systems that monitor dam and weir operations remotely [21].

The energy demand sector includes the agricultural, construction, industrial and household energy sectors [19]. For example, the percentage of households with electricity supply increased from 55 per cent in 2001 to more than 80 per cent in 2017 [22]. This scenario reflects the increase in household energy demand and the consequent water demand. The MoHUA's Smart City Mission aims to implement smart water solutions that collect real-time meaningful and actionable data from existing water networks [23]. Utilities can use this information to effectively distribute water. The mission's emphasis on artificial intelligence (AI), smart sensors, and technologies will improve leak detection by pinpointing leak locations, eliminate false leak alarms, enhance real-time monitoring of the network, and improve water quality issues and customer services [23]. An efficient pumping system is a key strategy to improve household water management. Emphasis on the reuse and recycling of wastewater in buildings should be supplemented by a decentralized water purification system at the city level.

To reduce the energy footprint of water and minimize wastage, Indian cities have begun to take some measures. The efficiency of water pumping systems is being improved in cities with appropriate rationalizing and pricing mechanisms. In Bengaluru, some apartment buildings have been built with smart water metering, which facilitates hourly water tracking and remote management of leaks [24]. The Indian Green Building Council's (IGBC) green cities rating system provides incentives to reduce water consumption and aid reduction by metering and monitoring water consumption. Alternative energy sources are also being utilized in high-water-consuming sectors. For example, solar energy is being used for electricity generation to ease the burden of water-intensive thermal power production processes.

In order to integrate SEM into the water sector, several factors need to be considered. Some of the challenges facing the water sector are the lack of proper metering, wherein the true cost of water prices is not calculated. An evaluation and water pricing mechanisms to measure the efficiency of the water systems (e.g., pumps) are rare. The automation of the water systems is very limited. The limited capacity of a household to heat water at any time (e.g., sun availability) is a challenge for renewable energy in the water sector. In addition, spatial, temporal and socio-economic changes and other political conditions can affect water availability. With the help of SEM practices, water and energy losses are likely to decrease and the efficiency of the water system can be improved.

3.3. Achieving Smart and Low-Carbon Mobility

Transport plays a pivotal role in the development of a country. A transportation network fosters passenger and freight movement across the country, thereby increasing national productivity and socio-economic growth. The increase in transport demand has made the transport sector one of the most energy- and carbon-intensive sectors in India. The transport sector accounts for 24 per cent of total energy consumption [25]. On the other hand, the sector accounts for 13.2 per cent of the total carbon emissions [26]. With growing concern about energy security and climate change, it is now recognized that the transport sector should reduce its reliance on fossil fuels, energy consumption and carbon footprint.

The energy consumption and carbon emissions of the transport sector are typically determined by factors such as vehicle efficiency, vehicle use and distance travelled, fuel and energy types, and overall system efficiency of transport infrastructure [27]. To promote energy-efficient and low-carbon growth in the road transport sector, the GoI has introduced several policies and programmes across passenger and freight segments. The main focus of the policies in this transport segment is on the improvement of vehicular technology through the implementation of progressive fuel efficiency norms and electrification.

With an objective to promote energy efficient low-carbon growth of the road transport segment, the government has introduced two major programmes: the Vehicle Fuel Efficiency Program and the National Electric Mobility Mission Plan (NEMMP) 2020. These are applicable for both passenger and freight road transport in India and are being implemented in a phased manner. Under the Vehicle Fuel Efficiency Program, the implementation of fuel economy standards is an effective regulatory instrument to reduce the average fuel consumption of vehicles. In 2017, the Ministry of Road Transport and Highways (MoRTH) came up with the first set of fuel economy norms for light duty vehicles (LDVs) in the passenger segment. These standards are based on the corporate average fuel economy (CAFE) norms and define the targets in terms of fuel consumption in liter/100 km.

Under the NEMMP, launched in 2013, the Faster Adoption and Manufacturing of (Hybrid and) Electric Vehicles (FAME) scheme was launched in 2015 by the Ministry of Heavy Industries and Public Enterprises to incentivize the production and promotion of electric vehicles (EVs). In addition to the private vehicle segment, the government has introduced EVs in multimodal public transport. In 2017, Nagpur became the first city in India to launch an electric mass transit project in India. A fleet of 200 electric vehicles (100 electric taxis and 100 e-rickshaws) was procured, and a cab aggregator provided the service platform for running the e-vehicles. Furthermore, several mobility solutions such as a public bicycle sharing scheme, intelligent transport management systems, electric feeders for last/first mile connectivity, integrated transport management platforms, and development of ICT applications have been proposed by several smart cities within the Smart Cities Mission.

The future of low-carbon transport should be highly efficient electric cars running on renewable electricity, a shift from private cars to public transport, better urban planning and investment in options that promote non-motorized transport (NMT) such as cycling and walking. In addition, the current electricity grid infrastructure will need to be reinforced if a significant level of transport electrification takes place. As the transport sector has implications on various other sectors, a cross-sectoral approach that incorporates reviewing the economic and environmental feasibility of sustainable mobility options should be undertaken. To achieve sustainable and low-carbon mobility in cities, the challenges highlighted in the discussions were the need for contextualized transport choices, informed decision making by policymakers, and sensitization of policy to citizens.

The recommendations suggested in the discussions included the promotions of cab aggregators/service providers such as Ola and Uber, provision of subsidies for EVs, and support for NMT options such as trams, as observed in various European countries. Cities also need to promote innovative solutions using ICT and efficient data and energy management. Policies for the transport sector should not be developed in isolation. Policy inputs from all sectors need to be taken into consideration.

3.4. Optimizing Waste Management Processes

From 2000 to 2015, the urban population of India almost doubled, while the amount of waste generated by the population increased by 2.5 times [28]. In addition, while the urban population has an annual growth rate of 3–3.5 per cent, urban waste generation is expected to increase by 5 per cent per year [29]. In this scenario, solid waste management is a major concern for cities. It is estimated that India's waste generation will reach 436 million tons by 2050 [30]. Effective waste management requires data management and integration at different levels, promoting the private sector and developing linkages between different sectors. Further, municipalities will need to focus on developing institutionalized and environment-friendly mechanisms to support proper waste disposal and better quality of life.

Intelligent energy management solutions that convert waste into useful energy can reduce the amount of waste generated and optimize the waste management process. A typical waste-to-energy (WtE) plant usually requires a minimum input of 300 TPD solid waste so as to make the system economically viable [16]. If large amounts of urban waste generated can be converted to energy, it can reduce the burden on conventional energy sources and the need for open space to dump unrecyclable waste. By 2050, India's WtE potential is estimated to become 556 megawatts (MW). However, these plants require diligence, adequate supply of quality waste, market infrastructure, and technical capacity [30]. Through proper support and the provision of smart technologies, municipalities can develop an active energy generation sector that has co-benefits for other sectors.

Effective technologies can be used for SEM in different areas of waste (collection, processing, and disposal). For example, radio frequency identification (RFID) technology, global positioning system (GPS) routing systems, and vacuum systems can reduce the time and effort spent on collection [31]. In waste treatment facilities, mechanical biological treatment and refuse-derived fuel (RDF) facilities ensure proper disposal of hazardous waste [31]. Moreover, sanitary landfills, bioreactor landfills, and solar integration mechanisms are treatment technologies that help convert excess waste into profitable energy [31].

The informal waste recycling industry is the entry point for introducing innovative and smart solutions. For example, intelligent recycling solutions ensure that informal waste sorting methods, such as manual rag-picking at landfills before the segregated waste is sent to recycling plants, are not only technically more advanced but also faster and safer to use for the workers. As a way forward, a smart waste management plan needs to be supported with the concept of circular economy.

According to Swachh Bharat Mission (Urban) data for 2018, 43 per cent of the total urban wards in India are now segregating their waste at the source [32]. In 2017, door-to-door collection coverage increased from 53 to 80 per cent [32]. In cities such as Panaji, Indore, Mysore and Muzaffarpur, there is a waste separation system, wherein separated waste is brought to the processing center [33]. Then, compost is made from wet waste, while only inert waste goes to the landfill. Sambyal [29] elaborates that Alappuzha in Kerala prioritizes segregation and reuse of waste at the household level, making it one of the cleanest cities in India.

It has accomplished decentralized waste management; 80 per cent of the households now own biogas plants and pipe composting systems. As part of the Clean Home Clean City programme, Alappuzha launched Thumburmuzhi in 2013, a model aerobic composting plant that composts animal carcasses [29].

A key challenge facing the waste sector in India is the need to increase manpower at the collection level. Waste segregation is an important obstacle and remains a daunting task. Despite the existing intelligent mapping and routing technologies, the segregation of waste, especially at the household level, is still limited. The sector requires a higher utilization of economical and user-friendly technical solutions.

Another core issue in the sector is the lack of accountability and transparency. Due to the limited knowledge of stakeholders (sometimes corruption) and the lack of innovative solutions, the methods used are not the optimal for effective waste management. Therefore, it is important to develop capacity building and awareness programmes for authorities and relevant citizens to respond to behavioural changes and incorporate smart practices into the waste management sector.

3.5. Enhancing Efficiency of Public Service Delivery

A range of urban public services such as street lighting, security management, video-surveillance, weather systems, and communication infrastructure provide safety, security, and information for citizens, while increasing the cities' competitiveness [34]. These public services need to be integrated with smarter, more energy efficient, and more innovative solutions for better service operations, management, and governance.

A range of urban public services, such as street lighting, security management, video surveillance, weather systems, and communications infrastructure, provide security, safety and information to citizens, while improving cities, combined with smarter, more energy-efficient, and more innovative solutions for better service operations, management, and governance.

SEM in public services helps city governments and utilities maintain and improve energy use, and to maximize the efficiency and quality of city services. Three different types of SEM solutions exist in the public services sector: (i) solutions that conserve, control, and monitor energy generated and distributed by utilities; (ii) solutions that store energy generated by customers or third-party members or the utility grid; and (iii) solutions that generate energy from natural resources, thereby creating relative or total energy independence from the grid [6].

A smart grid is an important technology for delivering utility-scale power to industrial, commercial, and residential areas in an efficient, reliable and safe way [6]. It consists of an independent energy network capable of exchanging electricity and operating systems in real time [6]. A micro/macro-scale smart grid can not only reduce energy loss, but also improve the utilization of renewable energy sources [35]. Smart substations and smart metering are the next steps in this direction.

Evidence suggests that smart substations [36] and advanced metering infrastructures [37] have improved the continuity of distributed supply and have had a positive impact on energy efficiency. Energy storage solutions (ESS) are used to store different types of energy (e.g., electricity, heat, kinetic energy). In urban public services, ESS can be used to integrate renewable energy and support demand-response plans. An important advantage of ESS is that customers or third-party energy producers can store energy from the utility grid during a lower price period and use it during a higher price period. Recent advances in energy storage technology include batteries, supercapacitors, flywheels, hydrogen fuel cells, compressed air storage, thermal storage, and mixed ESS [6,38]. Key applications of these technologies include battery-based grid systems, micro-grid and small-scale renewable energy technologies, and smart charging plug-ins for electric vehicles.

Finally, it is important to note that one of the goals of smart cities is to gradually migrate their electricity, thermal and data infrastructure to a complete renewable energy based systems [39]. Cities need to localize electricity consumption, provide low-carbon heating and cooling, and recycle energy and resources to maximize efficiency. Solutions that support this approach include solar photovoltaics (e.g., grid-connected, off-grid), solar collectors, centralized solar power plants, small and utility-scale wind turbines and geothermal energy. Other non-renewable resources with less impact, such as combined heat and power (CHP) and natural gas and biomass power generation, can better replace conventional power generation.

Local governments in India have been implementing smart energy strategies in public services. For example, district regional cooling systems, smart grids, smart metering, net metering and renewable energy integration are being planned or already used [16]. However, the lack of effective policies and regulations at the central, state and municipal levels, as well as inadequate guidelines, standards and business models, are obstacles to the large-scale use of widely available public service-based SEM technologies (e.g., energy storage, smart micro-grids). Integration of information modelling into urban management infrastructure (climate monitoring stations, lighting and power outage management controls, underground utility monitoring infrastructure, and data and communications management stations) is another key thrust area.

Following the recent announcement of 100 per cent electrification, GoI is making every effort to provide a reliable 24X7 of electricity to all its citizens and to promote reliable and transparent delivery mechanisms. In addition, with the launch of the Smart Cities Mission, hydropower, transportation, telecommunications and disaster management organizations are adopting the latest technologies to improve operational efficiency. Applying SEM solutions in public services can help achieve urban management goals through efficient distribution and transmission planning, utility transformation, and technology transformation [7].

Figure 1 presents a summary of key SEM solutions in different sectors, and Figure 2 presents the key challenges for SEM in different sectors.

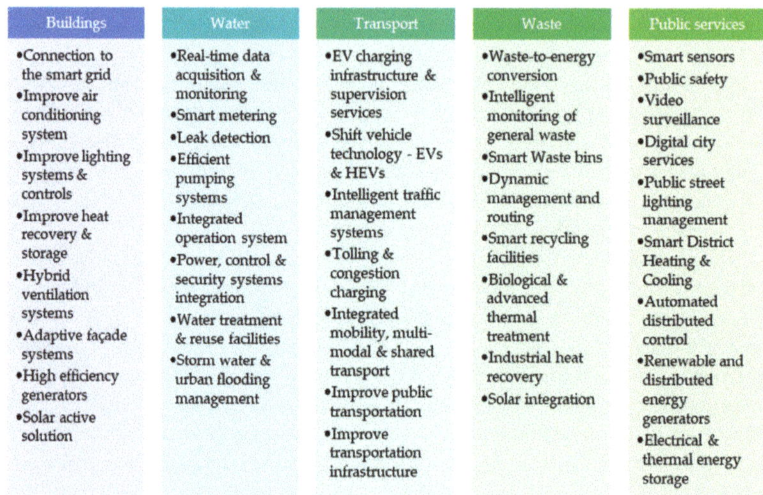

Figure 1. A summary of key smart energy management (SEM) solutions in different sectors.

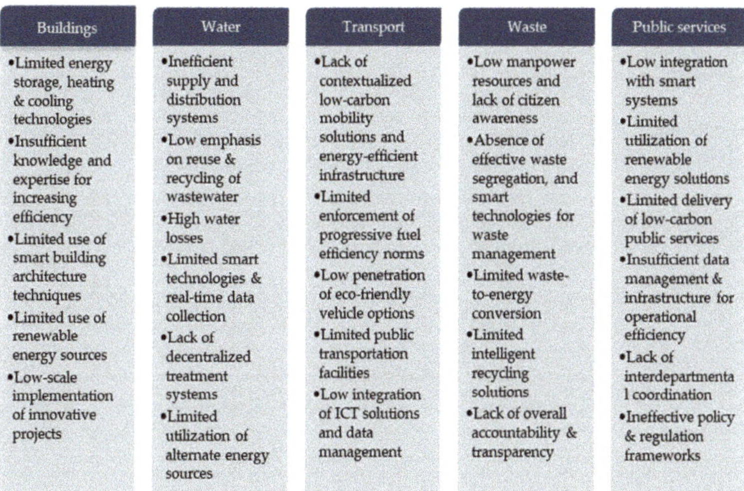

Figure 2. Challenges for SEM solutions in different sectors.

4. Overall Challenges and Opportunities for SEM in India

In addition to the above sectoral challenges (Figure 2), different sectors face a range of cross-cutting challenges. Overcoming these challenges is necessary to facilitate and accelerate the implementation of SEM projects. These challenges must therefore be identified in order to allocate efforts and resources effectively and to reduce the main obstacles.

A range of multidisciplinary policy, management and administration tools and techniques are available to analyze the viability of policy and process development of SEM. Examples include from simple policy analysis tools such as SWOT analysis and root cause analysis to more advanced multi-criteria decision making, metric approach, Kaizen 5S method, and Hoshin Kanri X matrix [40]. However, PASTEL analysis is used in this paper as it outlines a novel approach for addressing the political (P), administrative (A), socio-environment (S), technological (T), economic (E), and legal (L) challenges and barriers that constrain the development of SEM in India [41]. The PASTEL analysis listed in Table 1 contributes to a better understanding of the challenges and implementation barriers of SEM.

Table 1. PASTEL (Political, Administrative, Socio-environment, Technological, Economic, and Legal) analysis for SEM in India.

Area	Challenges
Policy	India lacks long-term and consistent SEM plans and policies. Despite, or perhaps because of, overlapping policies and complex urban governance arrangements, SEM governance in India remains fragmented and lacks political commitment and support on the long term.
Administrative	There is a lack of portfolio alignment between different sectors, and lack of good cooperation and acceptance amongst partners remains a major challenge. Different actors in the energy ecosystem may have competing aims and objectives. This poses a cross-sectoral design challenge. Long and complex procedures for authorization of project activities, complicated and non-comprehensive public procurement, difficulties in the coordination of high number of partners and authorities, and fragmented ownership are major barriers for scaling-up of projects. Public participation is rare and resources (institutional mechanisms, human, infrastructure, and skills) to disseminate information are limited.

Table 1. *Cont.*

Area	Challenges
Socio-Environment	Government led initiatives, such as demonstration and pilot projects, are needed for the majority of SEM actions in India. Negative effects of SEM related (e.g., solar and wind projects) project interventions on the social and natural environment may create inertia and interest in people.
Technology	Procurement businesses, skilled and trained personnel, and proven and tested solutions and examples are inadequate. Industry interest in SEM projects is limited, voluntary, and without strong influences, as the area is new and full of risks and planning deficiencies.
Economy	Limited access to capital and insufficient external financial support/funding for projects combined with economic crisis, risks and uncertainty in new technologies, and high costs of products and materials remain impediments for large-scale application of SEM solutions.
Legal/Regulatory	The extent of favorable and effective regulations and financial incentives for innovative and new technologies is insufficient. The lack of consistent regulations to standardize technologies is a major barrier.

There is a range of actions that governments and policymakers can promote to address energy-related challenges and achieve successful SEM in cities.

Integrated policy governance and effective decision-making—Energy management has traditionally been a part of either national or state government policy, while urban development and smart cities fall under the purview of state and city-level governments. Different stakeholders in different sectors with competing targets and goals may pose a significant SEM project and process design challenge. Therefore, the design process of SEM solutions needs cross-cutting initiatives. In addition, as the relation between energy and urban development becomes stronger, integration of SEM initiatives into all relevant government policies and operations becomes imperative. This should be supported by effective decision-making models. Multidisciplinary decision-making tools and techniques offer a variety of options and can facilitate the process of selecting the best solution within existing resources and support paradigm flexibility and applicability at any decision-making scale and variety [40]. Examples of existing decision-making methods for viable policy and process development include multi-criteria decision-making, process and content-oriented decision-making frameworks, decision-making matrixes, and qualitative decision-making tools [40].

Better governance will help the central, state governments and other stakeholders involved in different sectors to better coordinate to improve the effectiveness of energy management in smart cities-related policy decisions and public participation. SEM can be converged with the existing policy programmes by (i) establishing an inter-sectoral coordination committee that ensures integration, cross-referencing and liaison between appropriate organizations in buildings, transport, water and waste, and public services; and (ii) integration of SEM practices into relevant national and local policies (e.g., Smart Cities Mission, relevant missions under the NAPCC, urban development plans, building regulations, and procurement arrangements). Each of these actions can promote accountability and transparency in the decision-making process, which will contribute to smart energy governance in smart city development.

Provide better resources and infrastructure for technological advancements—Both human resources, and equipment, are required for adopting the intended SEM functions in smart cities. Funding and developing infrastructure for largescale applications of SEM initiatives remain a challenge. Therefore, governments must focus on innovative financing mechanisms and participation by both the public and private sectors. Governments should introduce adequate resources to drive informed decision-making, for investment prioritization in technology development, and to promote the scaling-up of SEM initiatives.

Policies and resources to support the ongoing development of new technologies are critical to facilitating the large-scale application of SEM initiatives in cities. For example, domestic research and development can reduce the relatively high import and capital costs, increase the potential sources of revenue for businesses and promote the viability of advanced SEM technologies [42]. Pilot and demonstration projects are important in proving the feasibility of new technologies. Successful demonstrations reduce the risk of investing in these technologies and help ensure private investment in large-scale projects [42]. Examples of technologies that have proven technical feasibility through small-scale demonstration projects include cogeneration, compressed air energy storage (CAES), and next-generation battery technologies such as sodium-sulfur batteries and liquid electrolytes low-cell-based batteries [43].

Develop information, education and communication (IEC) strategies for stakeholder awareness and engagement—Policies, actions and programmes that increase stakeholder engagement, induce behaviour change, encourage the adoption of smart energy solutions amongst the home and business stakeholders and increase education and awareness amongst the public, private sector and other stakeholders are necessary. Primarily, policies and initiatives that inform the public and stakeholders about the benefits of SEM can be implemented.

Some of the suggested measures are listed as follows:

1. Develop and implement a range of fiscal incentives, grants or subsidies, access to finance, tax breaks, and product rebates can be employed by governments to engage business, industry, and civil society.
2. Promote media campaigns (TV, radio, press, social media), printed materials (such as brochures, pamphlets, advertisements, posters, leaflets and electronic newsletters), national campaigns and competitions, conferences and events, online tools, and websites and exhibitions.
3. Engage business, industry 'champions', builders, planners, architects, transport networks, real estate organizations, energy consultants, and energy managers in enterprises in projects (e.g., community organizations in district cooling/heating projects) and outreach programmes.
4. Develop comprehensive guides, compendiums, data bases, and handbooks that bring together global and national best practices, examples, methodologies, technology solutions, and existing policies, measures, and programmes.

Establish performance goals for effective implementation and monitoring—The effectiveness of policy implementation depends on the performance outputs. To this end, governments should set a range of performance targets and measures to achieve the required outcomes. A long-term performance framework should be developed to ensure on-going SEM initiatives and intrinsically energy-efficient and -sufficient new assets. The framework could include specific responsibilities and obligations for agencies, accountable for managing sectoral policies, setting goals, monitoring performance and reporting against these goals, and measuring outcomes.

Performance objectives can be majorly of two types: (i) quantitative targets for energy use, GHG emissions, and renewable energy; and (ii) action-oriented targets (e.g., upgrading designated facilities and awareness-raising programmes). Performance objectives should be regularly monitored, analyzed and updated to meet the policy commitments. This information should be used as appropriate for annual reporting purposes and should include detailed information on each sector.

5. Conclusions

As India's urban population is expected to grow significantly over the next couple of decades, there will be huge implications for energy demand and consumption [2]. The task of reducing carbon emissions and transforming India's urban centres into energy efficient and sufficient cities requires integration of smart energy management (SEM) practices in the different sectors. SEM can contribute to sustainable and resilient energy systems and services by combining affordable and reliable technologies, active energy efficiency, energy conservation measures, and management of resources. This paper provides a comprehensive analysis of the policy structure and challenges and opportunities for SEM in India. Three main conclusions can be outlined from the review.

First, since the enactment of the Energy Conservation Act in 2001, a series of policy and regulatory reforms have evolved to support SEM in India. Specifically, the Jawaharlal Nehru National Solar Mission (JNNSM) and the Smart Cities Mission have been playing key roles in transforming India's energy management landscape, transforming the energy sector from its infancy into one of the world's largest policy markets.

Second, through the perspectives of different stakeholders involved in the Australia-India Knowledge Exchange Workshop in India and a comprehensive literature review, key SEM strategies and challenges in different sectors are discussed. In addition, some PASTEL (Policy, Administrative, Socio-Environment, Technology, Economy, and Legal) challenges are identified. In addition to reducing carbon emissions, improving energy security and increasing employment opportunities, addressing these challenges will help accelerate the spread of SEM projects and practices in India [41].

Third, a range of policy, governance, resources, information, education, and awareness-oriented initiatives were proposed to address the challenges of SEM. Governments should use each of these recommendations as a starting point; the recommendations are not intended to be prescriptive or exhaustive. The policy recommendations identified in this review are very useful to policy makers around the world who are interested in addressing the challenge of implementing SEM policies and ultimately supporting emission reduction targets.

Finally, SEM offers a bright future for India's energy and economic development. India could intensify its efforts to develop and implement SEM practices to achieve key energy, environmental and economic development goals. However, SEM can only be achieved through collaborative efforts between governments, practitioners, utilities, regulatory boards, and industries. SEM could change India's energy landscape: it has the potential to revive the economy by achieving energy independence, reducing the energy deficit and pushing it to become a "green nation".

Author Contributions: Conceptualization, K.Y. and R.R.; methodology, K.Y. and R.R.; formal analysis, K.Y., R.R. and A.P.; investigation, K.Y., R.R. and A.P.; resources, K.Y. and R.R.; data curation, K.Y., R.R. and A.P.; writing—original draft preparation, K.Y., R.R. and A.P.; writing—review and editing, K.Y. and G.P.; visualization, K.Y., R.R. and A.P.; supervision, K.Y and G.P.; project administration, K.Y.; funding acquisition, K.Y.

Funding: This research was funded by UNSW-India Seed Grant, grant number RG181589.

Acknowledgments: We would like to thank the two reviewers for their comments and suggestions. The reviews have significantly helped to improve the paper.

Conflicts of Interest: The authors declare no conflicts of interest.

References

1. IEA. *India Energy Outlook*; International Energy Agency (OECD/IEA): Paris, France, 2015.
2. UN DESA. United Nations, Department of Economic and Social Affairs, Population Division. 2018. Available online: https://population.un.org/wup/Country-Profiles/ (accessed on 22 March 2019).
3. Danish, M.S.S.; Sabory, N.R.; Ershad, A.M.; Danish, S.M.S.; Yona, A.; Senjyu, T. Sustainable architecture and urban planning trough exploitation of renewable energy. *Int. J. Sustain. Green Energy* **2017**, *6*, 1–7.
4. Cajot, S.; Peter, M.; Bahu, J.M.; Koch, A.; Maréchal, F. Energy planning in the urban context: Challenges and perspectives. *Energy Procedia* **2015**, *78*, 3366–3371. [CrossRef]
5. Energy Lens. The What, Why, and How of Energy Management. 2018. Available online: https://www.energylens.com/articles/energy-management (accessed on 21 March 2019).
6. Calvillo, C.F.; Sánchez-Miralles, A.; Villar, J. Energy management and planning in smart cities. *Renew. Sustain. Energy Rev.* **2016**, *55*, 273–287. [CrossRef]
7. Mosannenzadeh, F.; Bisello, A.; Vaccaro, R.; D'Alonzo, V.; Hunter, G.W.; Vettorato, D. Smart energy city development: A story told by urban planners. *Cities* **2017**, *64*, 54–65. [CrossRef]
8. MoEFCC. India's Intended Nationally Determined Contribution-Towards Climate Justice. In *The Ministry of Environment, Forest and Climate Change*; Government of India: New Delhi, India, 2015.

9. Rattani, V. *Coping with Climate Change: An Analysis of India's National Action Plan on Climate Change*; Mathur, T., Ed.; Centre for Science and Environment (CSE): New Delhi, India, 2018.

10. AEEE. The Need for a National Cooling Action Plan. 2018. Available online: https://www.aeee.in/national-cooling-action-plan/ (accessed on 20 March 2019).

11. Chedwal, R.; Mathur, J.; Agarwal, G.D.; Dhaka, S. Energy saving potential through Energy Conservation Building Code and advance energy efficiency measures in hotel buildings of Jaipur City, India. *Energy Build.* **2015**, *92*, 282–295. [CrossRef]

12. Danish, M.S.S.; Senjyu, T.; Ibrahimi, A.M.; Ahmadi, M. Abdul Motin Howlader A managed framework for energy-efficient building. *J. Build. Eng.* **2019**, *21*, 120–128. [CrossRef]

13. Karnouskos, S. Demand side management via prosumer interactions in a smart city energy market place. In Proceedings of the 2nd IEEEPES International Conference and Exhibition on Innovative Smart Grid Technologies, Manchester, UK, 5–7 December 2011.

14. Guan, X.; Xu, Z.; Jia, Q. Energy-Efficient Buildings Facilitated by Microgrid. *IEEE Trans. Smart Grid* **2010**, *1*, 243–252. [CrossRef]

15. Pulselli, R.M.; Simoncini, E.; Marchettini, N. Energy and emergy based cost–benefit evaluation of building envelopes relative to geographical location and climate. *Build. Environ.* **2009**, *44*, 920–928. [CrossRef]

16. Arunkumar, P.K.; Malur, R.R. *Energy Conservation in Built Environment in Smart Cities in View Point*; Consulting Engineers Association of India: New Delhi, India, 2018; p. 13.

17. Rana, N.P.; Luthra, S.; Mangla, S.K.; Islam, R.; Roderick, S.; Dwivedi, Y.K. Barriers to the development of smart cities in Indian context. *Inf. Syst. Front.* **2019**, *21*, 503–525. [CrossRef]

18. Bhutta, F.M. Application of smart energy technologies in building sector—future prospects. In Proceedings of the 2017 International Conference on Energy Conservation and Efficiency (ICECE), Lahore, Pakistan, 20–23 November 2017.

19. TERI. *Study on Assessment of Water Foot Prints of India's Long Term Energy Scenarios*; Niti Aayog; Government of India and The Energy and Resources Institute: New Delhi, India, 2017.

20. Development Asia. What Is Smart Water Management? 2018. Available online: https://development.asia/explainer/what-smart-water-management (accessed on 23 March 2019).

21. Development Asia. Sustainable Water Management for Smart Cities. 2017. Available online: https://development.asia/case-study/sustainable-water-management-smart-cities (accessed on 23 March 2019).

22. CPR and Prayas. Trends in India's Residential Electricity Consumption. 2017. Available online: http://cprindia.org/news/6519 (accessed on 23 March 2019).

23. Vaidya, A. Effective Water Management–Indispensable for Smart Cities. 2018. Available online: http://bwsmartcities.businessworld.in/article/Effective-Water-Management-Indispensable-for-smart-cities-/09-05-2018-148681/ (accessed on 22 March 2019).

24. Raj, A. IoT-Enabled Smart Meter Helps Hourly Water Tracking at Apartments. 2018. Available online: https://timesofindia.indiatimes.com/city/bengaluru/iot-enabled-smart-meter-helps-hourly-water-tracking-at-apartments/articleshow/62500517.cms (accessed on 22 March 2019).

25. TERI. *TERI Energy & Environment Data Diary and Yearbook 2016/17*; The Energy and Resources Institute: New Delhi, India, 2018.

26. UIC/IEA. *Railway Handbook on Energy Consumption and CO2 Emissions*; International Union of Railways & International Energy Agency: Paris, France, 2016.

27. Gulati, V. *National Electric Mobility Mission Plan (NEMMP) 2020*; Department of Heavy Industry, Government of India: New Delhi, India, 2012.

28. Sacchdeva, M. How IoT Enabled Smart City Helps Tackle the Problem of Solid Waste Management in India. 2018. Available online: https://www.dqindia.com/iot-enabled-smart-city-helps-tackle-problem-solid-waste-managementindia/ (accessed on 22 March 2019).

29. Sambyal, S.S. Waste Smart Cities. 2016. Available online: https://www.downtoearth.org.in/coverage/waste/waste-smart-cities-54119 (accessed on 21 March 2019).

30. NIUA & BEE. *Sustainable Energy Integration in Smart Cities: Background Paper*; National Institute of Urban Affairs & Blue Earth Enterprise: New Delhi, India, 2015.

31. Research Nester. Smart Waste Management Market. By Technology (Smart Collection, Smart Processing, Smart Energy Recovery) & By Geography 2016–2023. 2018. Available online: https://www.reuters.com/brandfeatures/venture-capital/article?id=59192 (accessed on 16 March 2019).

32. Sambyal, S.S.; Agarwal, R. Is Swachh Bharat Mission Ensuring Waste Segregation Systems? 2018. Available online: https://www.downtoearth.org.in/blog/waste/is-swachh-bharat-mission-ensuring-waste-segregationsystems-61885 (accessed on 22 March 2019).

33. EMBARQ Network. Four Cities' Solutions to Sustainable Garbage Processing. 2019. Available online: https://www.smartcitiesdive.com/ex/sustainablecitiescollective/friday-fun-how-create-tomorrow-s-green-cities-today-s-garbage/1050616/ (accessed on 16 March 2019).

34. Belanche, D.; Casaló, L.V.; Orús, C. City Attachment and Use of Urban Services: Benefits for Smart Cities. *Cities* **2016**, *50*, 75–81. [CrossRef]

35. Hossain, M.S.; Madlool, N.A.; Rahim, N.A.; Selvaraj, J.; Pandey, A.K.; Khan, A.F. Role of smart grid in renewable energy: An overview. *Renew. Sustain. Energy Rev.* **2016**, *60*, 1168–1184. [CrossRef]

36. Huang, Q.; Jing, S.; Li, J.; Cai, D.; Wu, J.; Zhen, W. Smart substation: State of the art and future development. *IEEE Trans. Power Deliv.* **2017**, *32*, 1098–1105. [CrossRef]

37. Sharma, K.; Saini, L.M. Performance analysis of smart metering for smart grid: An. overview. *Renew. Sustain. Energy Rev.* **2015**, *49*, 720–735. [CrossRef]

38. Dubal, D.P.; Ayyad, O.; Ruiz, V.; Gomez-Romero, P. Hybrid Energy Storage: The Merging of Battery and Supercapacitor Chemistries. *Chem. Soc. Rev.* **2015**, *44*, 1777–1790. [CrossRef] [PubMed]

39. Mathiesen, B.V.; Lund, H.; Connolly, D.; Wenzel, H.; Østergaard, P.A.; Möller, B.; Nielsen, S.; Ridjan, I.; Karnøe, P.; Sperling, K.; et al. Smart Energy Systems for coherent 100% renewable energy and transport solutions. *Appl. Energy* **2015**, *145*, 139–154. [CrossRef]

40. Danish, M.M.S.; Senjyu, T.; Zaheb, H.; Sabory, N.R.; Ibrahimi, A.M.; Matayoshi, H. A novel transdisciplinary paradigm for municipal solid waste to energy. *J. Clean. Prod.* **2019**, *233*, 880–892. [CrossRef]

41. Yenneti, K. The grid-connected solar energy in India: Structures and challenges. *Energy Strategy Rev.* **2016**, *11*, 41–51. [CrossRef]

42. Zame, K.K.; Brehm, C.A.; Nitica, A.T.; Richard, C.L.; Schweitzer, G.D., III. Smart grid and energy storage: Policy recommendations. *Renew. Sustain. Energy Rev.* **2018**, *82*, 1646–1654. [CrossRef]

43. Sioshansi, R.; Denholm, P.; Jenkin, T. Market and policy barriers to deployment of energy storage. *Econ. Energy Environ. Policy* **2012**, *1*, 47–64. [CrossRef]

MDPI

St. Alban-Anlage 66

4052 Basel

Switzerland

Tel. +41 61 683 77 34

Fax +41 61 302 89 18

www.mdpi.com

Energies Editorial Office

E-mail: energies@mdpi.com

www.mdpi.com/journal/energies